2990

The book of
INDIAN REPTILES

The book of INDIAN REPTILES

J. C. DANIEL

Curator, Bombay Natural History Society

With 41 coloured and 13 monochrome plates and numerous text-figures

Sponsored by the

Department of Science & Technology, Government of India

BOMBAY NATURAL HISTORY SOCIETY

OXFORD UNIVERSITY PRESS

BOMBAY DELHI CALCUTTA MADRAS

Oxford University Press, Walton Street, Oxford OX2 6DP

Oxford New York Toronto
Delhi Bombay Calcutta Madras Karachi
Petaling Jaya Singapore Hong Kong Tokyo
Nairobi Dar es Salaam
Melbourne Auckland
and associates in
Berlin Ibadan

First Published 1983
First Reprinted 1989
Second Reprinted 1992

ISBN 0 19 562168 9

PRINTED BY BRO. PAULINUS AT ST. FRANCIS INDUSTRIAL TRAINING INSTITUTE, MOUNT POINSUR, BORIVLI (WEST), BOMBAY 400103, PUBLISHED BY THE BOMBAY NATURAL HISTORY SOCIETY AND CO-PUBLISHED BY S. K. MOOKERJEE, OXFORD UNIVERSITY PRESS, OXFORD HOUSE, APOLLO BUNDER, BOMBAY 400039.

PREFACE

This book describes the natural history of reptiles commonly occurring in the Indian subcontinent, particularly within the political boundaries of the Republic of India. Brief descriptions of forms unlikely to be seen by those who have only a casual interest in these animals are also included for the sake of completeness, but no claim is made that this is exhaustive. Several excellent books on reptile biology are now available which synthesize the great strides made in the study of reptile populations, behaviour, morphology and other facets. Those which could be consulted profitably by the student of herpetology are listed at the end of this volume. It is not proposed to repeat here information that is already available. However, information on some basic features essential in any introductory volume is included. Considerable work still remains to be done even on the commoner Indian species.

The book is based largely on data available in Society's *Journal* fleshed out with personal observations. The numbers in parenthesis following any particular observation in the body of the text refer to the source reference listed at the end of the volume. The section on snakes is an up-dated and condensed version of the excellent series on Common Indian Snakes by Col. Frank Wall serialised in the early issues of the Society's *Journal.*

I am deeply grateful to the many members and the staff of the Society who gave photographs, namely Mr M. Krishnan, Mr Romulus Whitaker, Mr S. R. Sane, Dr Edward Moll, Miss Vijaya, Mr Isaac Kehimkar, Mr S. A. Hussain, Dr. A. R. Rahmani, Mr. B. C. Chowdhury, Mr. Helmut Denzau, Ms. Gertrud Newmann and Mr. S. R. Nayak.

I am also grateful to the Govt. of India and Dr Ananthakrishnan, formerly the Director, Zoological Survey of India, for permission to reproduce line drawings from the reptile volumes in the FAUNA OF INDIA by Malcolm Smith.

The colour plates of snakes are those which accompanied Col. Wall's series in the Society's *Journal.* The line drawings have been copied by Mr A. Gnanasekar, Research Assistant at the Society.

I am indebted to Dr Salim Ali, the late Mr D. E. Reuben, Dr Robert Bustard and Mr Romulus Whitaker for reading through and commenting on the manuscript of the book, and Mr J. S. Serrao and Mr B. Ganesan for the preparation of the typescript. The assistance of Mr J. S. Serrao and Mr K. P. Shirodkar of the Society's publication section and Br. Paulinus of St Francis Technical School was invaluable in seeing the book through the press.

J. C. DANIEL

CONTENTS

INTRODUCTION

Among the four classes of terrestrial or land dwelling vertebrates or animals with backbone, the reptiles form the middle element. They have evolved from the amphibians, whose present day representatives are the frogs, toads, newts and caecilians and they have given rise to the birds and mammals. As a group the reptiles can be separated from the amphibians by their dry scaly skin, from the birds by the absence of feathers, and from the mammals by the absence of fur or hair. However, such a simple definition of a reptile may not be adequate to describe reptiles such as the mud turtles which have very few scales to distinguish them from an amphibian. The breeding habits, that is the laying of eggs on land and the nature of the eggs which is described below is the character that distinctly separates reptiles from the water dwelling and water bound fishes and amphibians. All land animals that have scales are not necessarily reptiles. The majority of birds have scales on their legs but scales occur in association with feathers. Similarly there are mammals like the scaly ant-eater or pangolin fully covered in scales but fur or hair is always present.

The history of reptiles goes back many millions of years. On the basis of fossil evidence, they are believed to have originated during the geologic period known as the upper Carboniferous or Pennsylvanian period 300 to 260 million years ago. Reptiles have existed in some form or other since then. They were the dominant form of vertebrate life on earth for the following 140 to 120 million years, the Mesozoic era, of earth history. This was the age of reptiles, the period when the dinosaurs flourished, giant reptiles whose variety, size and bizarreness has excited the imagination of man and left him with a lasting regret that none survived to his time and age. Conan Doyle's novel the *Lost World* and the odd rumours of the presence of a large reptile in the swamps of central Africa are an expression of this regret. The largest land animals lived during this age, epitomized by the giant herbivore, the *Brachiosaurus* with an estimated weight of about 78 tonnes and the giant Carnivore *Tyranosaurus rex* which towered to a height of over 6 m and attained a weight of 8 tonnes.

Modern reptiles, which appeared during the Tertiary period of earth history some 70 million years ago, are but a fraction of the number and variety that lived during the age of reptiles. Even presently, the number of species of living reptiles, which is about 6000 is almost double the number of species of present day mammals. Half the number of living reptiles consists of lizard species.

The conquest of land by reptiles, the first truly land vertebrates, was made possible by the evolutionary development of four embryonic membranes which provide the embryo its own aquatic environment in which to develop. A calcareous or parchment-like egg shell gives protection. The four membranes are the Chorion, or outer sac, immediately inside the shell which surrounds and gives protection to the whole egg; the Amnion, the inner sac, which filled with fluid provides the aquatic environment for the developing embryo; the

Allantois, which develops from the hindgut of the embryo to provide for respiration and storage of waste products; and the Yolk sac, storing the food for the developing embryo. The embryo is cushioned against shock by an albuminous liquid the final protection against the hazards of starting life on land. Thus vertebrate life permanently broke its link with water.

Reptiles used to be termed cold blooded, meaning that their body temperature varies with the outside temperature whereas birds and mammals which have a constant body temperature were known as 'warm blooded'. Reptiles have a low metabolic rate and therefore produce less heat than a mammal or bird of comparable size. They have poor body insulation and cooling mechanism as they lack sweat glands yet they have considerable capacity for regulating their body temperature. Thermo-regulation in reptiles is a behaviour function and is achieved by judicious use of available sunlight. By basking in the sun or absorbing heat through a hot substratum when heat is required and moving away from the sun when heat is not needed, reptiles are able to maintain an ideal temperature status within their body, which may be more or less than the air temperature of their habitat. The majority of reptiles being inhabitants of the tropics, temperature regulation presents no problem. Those living in temperate climates hibernate during the winter months. Similarly very hot dry seasons in the tropics are tided over by aestivation. The reptiles by the very nature of their temperature control behaviour are able to survive only within a narrow band of environmental temperature and even within this band operate best at species-specific preferred temperatures which lie between 20 to 38°C (68 to 102°F) depending on the species.The Tuatara of New Zealand, geologically the oldest of living reptiles, is an exception having a preferred temperature of 12°C (54°F). Reptiles once they have achieved optimum temperature remain inactive for long periods except for brief bursts of activity for bodily functions such as feeding, sex and for safety. Sustained activity similar to that of birds and mammals is beyond most reptiles, the exception being sea turtles which migrate long distances.

Shape and Size: Three types of body form are seen among reptiles. The basic type is the lizard-like shape. In this form, ground, wall and tree trunk dwelling reptiles are flattened dorso-ventrally which facilitates concealment and movement. The crocodiles, monitor lizards and geckos are examples of this type of body shape. The legs are well developed. In arboreal forms the body is flattened laterally as in Chameleon and the garden lizards, the Calotes. In the second type, the body is elongated and cylindrical as in many skinks and all the snakes; legs may be rudimentary or absent. In the third type, the turtles and tortoises, the trunk has become rigid and enclosed in a bony shell.

Indian reptiles range in size from the massive estuarine crocodile with lengths up to and perhaps exceeding 7 m and geckos less than 100 mm in total length.

Locomotion: The movement on land and water is dictated by the shape and size of the animal. Quadruped or four legged reptiles have what could be termed a 'low walk', the body being slung like a palanquin between the legs and not carried high on the legs as in mammals. An exception is the Crocodile, which can walk high on its legs. Juvenile crocodiles also have the ability to gallop. The running gait of most quadruped reptiles is a trot, the front and back legs being moved diagonally. From the nature of suspension of the body between the legs, long bodied reptiles like the monitor lizard have, while running, a side to side movement on the long axis, as the body curves with the movement of the hind legs. Some species like the Sitana or Fanthroated lizard when hard-pressed, run bipedally for short distances.

Lizards and Snakes are the only reptiles that can climb and many are arboreal. The Chameleon and some of the Geckos being particularly adapted for arboreal life. One Gecko, one lizard and one snake have the ability to glide through the air.

The characteristic locomotion of snakes, termed serpentine, a method of progression by lateral undulations of the body, is also the mode of locomotion used by skinks and other reptiles having an elongated snake-like body with limbs vestigial or absent. The muscular development of the trunk permits sideways looping of the body enabling the animal to make use of the uneveness of the ground to move forward. On smooth sandy or similar surfaces some species such as the Phoorsa (*Echis carinata*) move by a method known as side winding, the animal being parallel to instead of facing the direction of movement and loops being lifted off the ground and thrown forward. Some of the larger snakes like the python can move with the body in a straight line, albeit very slowly.

Senses: The three major senses of sight, hearing and smell show varying degrees of development in reptiles. In addition snakes have special sense organs, the sensory pits on the snout.

Sight: The eyes are well developed in most reptiles except the burrowing forms. In most diurnal reptiles the pupil is circular, the majority of nocturnal reptiles, like Geckos, and Crocodiles, have a vertical pupil; the Whip Snake's pupil is peculiar in being horizontal probably connected with the binocular vision possessed by these snakes. All snakes, and Geckos among lizards, do not have eyelids but have a permanent transparent cover changed every time the skin is shed. Many skinks have a transparent window on their lower eyelid permitting vision with the eyelids closed. The night vision capacity of the crocodiles is increased by the presence of a special cell layer called tapetum within the eye which reflects light and increases light usage. The tapetum layer makes crocodile eyes reflect at night with a red glow in torchlight.

Hearing: The sense of hearing is well developed in the crocodilians which are provided with an external ear flap to shut off the tympanic cavity when under water. In other reptiles when the ear is present the tympanum is flush with the surface or in a slightly sunk cavity. An external ear is absent. Snakes and some of the burrowing lizards lack the middle ear also and their hearing is mainly restricted to vibrations carried through the substratum, the lower jaw being the main conductor. Reaction to sound when once alerted is mainly visual as a snake above ground or on a tree moves with the head raised and sound reception through the ground is then reduced in effectiveness. Generally the sense of hearing is of low efficiency in reptiles.

Smell: Reptiles, except the crocodiles and some of the arboreal lizards, have an additional organ of smell, Jacobson's organ, which is indeed the main organ of smell in the snakes. In snakes, the forked tongue has lost the traditional function of the tongue as an organ of taste and has become the carrier of scent particles, to the ducts of Jacobson's organ. The snake smells its food and its enemies with its flickering tongue. In other reptiles the importance of the nasal passage and Jacobson's organ vary. In crocodiles the sense of smell has remained with the nostrils. The acuteness of the sense of smell of a reptile varies with its life habits.

Sensory pits: A remarkable sense organ exclusive to snakes is the labial or lip pits seen in the Boid snakes such as the Python and the paired facial pits seen in the subfamily Crotalinae or Pit Vipers. The facial pits of the vipers are more sophisticated in structure than the lip pits of the pythons, and have a thin highly innervated membrane dividing the pit into an inner and outer chamber. Both types of pits are very sensitive to heat radiation and assist in the location and capture of warm-blooded prey.

The brain in reptiles is small, about equivalent in size to those of similar-sized fish or amphibians. Reptiles have low intelligence and are ruled mainly by inborn or instinctive behaviour and lack the flexibility and adaptability seen in mammals. Even in the quality of instinctive behaviour they are far inferior to birds.

Food and feeding habits: The majority of reptiles are carnivorous the snakes being exclusively so. The animals eaten range in size from an occasional large mammal in the case of a crocodile or python to a variety of insects in the case of the majority of lizards. Insects are, in fact, the main food of a large number of reptiles as diverse as young crocodiles to young snakes. A few reptiles such as the land tortoises, the marine green turtle, and the spiny tailed lizard are partially or totally herbivorous. Many species are adapted to special diets like the egg-eating snakes and some such as the king cobra, are cannibalistic. In their choice of food reptiles as a group are of great benefit to man from the number of insects and rodent pests eaten by them.

Reptiles are not active predators, in general taking prey which their immobility makes unwary; but the monitor lizards, the skinks and a large number of snakes actively search for prey. Normally only the mouth is used in the capture of prey which, if it is large, is subdued by being battered on the ground, or choked to death in the coils of the body as is the case with the python and other Boid snakes or killed by poison by the poisonous snakes. Snakes and many lizards have recurving teeth useful in the capture and holding of prey. Most reptiles replace their teeth throughout life, one will never meet a toothless crocodile or a toothless snake. Teeth may change in shape with age and type of prey. Only the turtles and tortoises lack teeth, these being replaced by a horny beak which is, in some species, serrated. The tongue's traditional function as an organ of taste probably remains only in lizards, crocodiles and tortoises. In the latter two it cannot be protruded out of the mouth. The tongue is the organ used to capture prey by the Chameleon and has become a slender, sensitive, organ of smell in snakes.

Poison in reptiles is a means of capturing prey and a method of defence. Experimental evidence indicates that in the great majority of snakes, the secretion from the parotid or cheek glands which flows into the mouth is poisonous particularly to their prey species. However as far as man is concerned the quantity of poison secreted by most species of snakes is so minimal in a single injection that they are harmless. Only the snakes which secrete and store poison in quantity are fatally dangerous to animals larger in size than their normal prey. However, gravity of symptoms resulting from a snake bite is entirely dependent on the quantity of poison injected and absorbed by the system.

None of the Indian lizards are poisonous. The cases reported of people falling sick after eating liquid or semi-liquid food in which geckos have accidentally drowned are the result of acute food poisoning of bacterial origin.

Reproduction: As the first truly terrestrial animals, the reptiles introduced innovations in their reproduction strategy. For example the fertilization of the egg within the body of the female by the act of copulation ensures that the sperm which can live and travel only in a fluid medium, is fully protected against desiccation in the transfer from male to female. The protection of the membrane covered egg by a shell provides for the safe development of the egg on land.

Females of many reptiles have the ability to store sperm and fertile eggs may be laid even three to four years after separation from the male.

The majority of reptiles lay eggs which are usually buried in pits in the soil for incubation. The saltwater crocodile and the king cobra lay in specially constructed nests above ground. Parental care is normally an exception but is seen among the crocodiles, pythons, cobras and some of the skinks. The female python incubates its eggs.

Viviparity or live birth in which the eggs develop within the mother, is a feature of reptiles which live in cold environments and of the sea snakes which are thoroughly adapted to an aquatic existence and are unable to survive on land. Viviparity may be in the form of ovo-viviparity in which the eggs incubate within the mother and often shelled unfertile eggs are voided along with the fully formed young at birth; or the mechanism may be truly viviparous in that a placental connection exists between the developing embryo and the mother.

Courtship among reptiles is not as elaborate as among birds and is based on male dominance.

Distribution of Reptiles: The distribution of animals or their zoogeography gives indication of not only their origin and environmental needs but also the history in time of continents and their fauna. There is evidence that 300 millions years ago there was a single continent, Panagea, which split into the world as we know it today as the component underlying continental plates drifted apart. The theory of continental drift is substantiated by the distribution of primitive forms of animal life, for instance, the limbless amphibians of the family Caecilidae which occur in South America, Africa, India and countries in southeast Asia up to the Philippines. India with South America and Africa formed a single large continent, Gondwana, before separation into different continents, and many of the older genera of Indian reptiles have species in the African fauna.

At the present time six zoogeographical regions are recognized, each with the majority of its fauna peculiarly its own. The zoogeographic region in which India is situated, the Oriental Region is bounded on the west and north by the Palaearctic European faunal region and on the southeast by the Australian region.

The Oriental region has three main faunal subregions. The Indian subregion closely follows the present political boundary of India in the Indian subcontinent and includes Nepal, Bangladesh and Sri Lanka. The Indochinese subregion consists of the Himalayas mainly east of the Arun river, the states of eastern India, Andamans, Burma, Thailand and Indochina. The Malayan subregion includes the Malay Peninsula and islands to the east. The Nicobar group is predominantly Malayan in its faunal affinity.

The reptilian fauna of India is largely dominated by Indochinese elements, relicts of a period in India's geologic history when the Peninsula had much more rainfall than what it experiences today and the vegetation and ecological factors were identical to what obtains now in the Indochinese faunal section of the country, the eastern Himalayas and Assam and other states of eastern India. A relict of this fauna now occurs in the high rainfall area in the Western Ghats principally below 15°N latitude. Several genera such as *Draco* (Flying Lizards), *Chrysopelia* (Golden Tree Snake), *Heosemys* (Forest Terrapin),

Gharial

Hawksbill Turtle *Courtesy of* WWF/Urs Woy (see pp. 15, 21)

Photo: T. N. A. Perumal

Mugger or Marsh Crocodile.

Photo: M. Krishnan

Estuarine Crocodile (See pp. 10, 13)

Dasia, Lygosoma, Leiolopisma (Skinks) have the same species or closely related species occurring in the Western Ghats and eastern India with an intervening gap in distribution in the Peninsula where conditions have now become unsuitable for these high rainfall habitat species. Among the 140 genera of reptiles that occur in India as a political entity, 67 occur in the Western Ghats and 56 in eastern India. Among these *Dravidogecko* (Gecko), *Salea* (Agamid Lizards), and *Ristella* (Skink) are endemic to the Western Ghats. The family Uropeltidae, primitive burrowing snakes are exclusive to the Western Ghats and Sri Lanka. Thirteen genera of reptiles occur only in eastern India and have not penetrated further into the Peninsula. The fauna of the Nicobar group of islands show their Malayan affinity in the presence of such genera as *Ptychozoon* (Gliding Gecko) and the family Dibamidae (Limbless Lizard). The distribution of the Emerald Geckos (*Phelsuma*) is peculiar. Apart from the Andamans, species of this genus occur only in other oceanic islands such as the Seychelles and are probably true relicts of the fauna of Gondwanaland.

Animals which have a wide distribution have adaptability and tolerance to a wide variety of environmental conditions and could be considered as the most successful species of a group. Among Indian reptiles, species having a wide distribution in almost all biotopes except those with the most extreme of environmental conditions are: the Mugger Crocodile *(Crocodylus palustris)*, Common Mud Turtle *(Lissemys punctata)*, Brooks's Gecko *(Hemidactylus brooki)*, Common garden lizard *(Calotes versicolor)*, Little Skink *(Mabuya macularia)*, Common monitor *(Varanus bengalensis)*, Common Blind Snake *(Typhlina bramina)*, Indian Python *(Python molurus)*, Dhaman or Rat Snake *(Ptyas mucosus)*, Common Wolf Snake *(Lycodon aulicus)*, Chequered Keelback *(Xenochrophis piscator)* and the three harmful snakes Common Krait (*Bungarus caeruleus*), Cobra *(Naja naja)* and Russell's Viper *(Vipera russelli)*.

Man and Reptiles: The majority of reptiles are useful and do silent service to man in controlling agricultural pests both insects and rodents but excite little interest in man. If it were not for the harmful snakes, reptiles as a group would be largely ignored. The greatest danger to this useful group of animals is from the demand for their skin for commercial purposes. Malcolm Smith drew attention nearly fifty years ago to the danger of extinction facing many reptile species from this cause. The skin trade has made Indian crocodiles endangered and the demand for snake and lizard skins remains a constant threat to the survival of such useful reptiles as the Monitor Lizard and the Rat Snake.

CROCODILES

Crocodiles (Family Crocodylidae) belong to an ancient group of reptiles in existence for millions of years. Crocodilians very similar to present day species appeared on the earth during the Triassic era of earth history, *c.* 190 million years ago. The survival of the crocodilians over a long period of earth history is perhaps due to their needs being easily met, living as they do on the edge of two life zones, water and land, and being able to find their prey from both zones. Indian crocodiles look alike and it is difficult to distinguish in the wild, without considerable experience, the Marsh Crocodile which lives in the lakes and rivers of India and the Estuarine Crocodile which lives in the coastal river estuaries. Only the Gharial, or Long-snouted Crocodile, of the great rivers, the Indus, the Ganges, the Brahmaputra and the Mahanadi can be easily identified by its comparatively longer and slender snout. All crocodilians show certain structural adaptations for a successful aquatic life. The nostril is placed at tip of the snout enabling the animal to breathe when the rest of the body is submerged. The nostril and the ears are provided with flaps for closing them when the animal submerges. The eye has a transparent third eyelid permitting limited underwater vision. Folds on the tongue and palate prevent water from entering the lungs when the mouth is open under water. Curiously enough the crocodile cannot protrude its tongue like other reptiles.

The shape of the body follows the generalised reptilian pattern of long body and tail with short limbs. The skin of the back is armoured with bony plates arranged in transverse series. The tail, similarly armoured, bears two rows of serrated scales which merge before the tail end.

The long, and in the case of the Gharial, slender snout is armed with conical teeth set in bony sockets on the edge of the jaws. The true crocodiles, can be separated from the alligators by the arrangement of the teeth. In the crocodiles the fourth tooth of the lower jaw fits into a notch on the side of the upper jaw and is visible when the mouth is closed. In the alligator, this tooth is contained within the mouth and does not show externally. Alligators do not occur in India. As in other reptiles, crocodilian teeth are shed and replaced throughout life.

The senses of sight, smell, and hearing are well developed. The nostrils, placed near the tip of the snout, are connected by long tubes to the throat behind a dividing flap from the mouth, permitting breathing with the prey in the mouth while partially submerged. The vent or cloaca has a longitudinal opening. The male has a single penis situated within the cloaca. Two pairs of scent glands are present, one on the throat and the other within the lips of the cloaca.

Crocodilians are excellent swimmers, the tail being the main propellent. On land they walk or slide on their belly. Crocodiles have what is termed a 'high walk' with the body raised well above the ground, at a speed of about 3 km per hour. The ability to gallop to escape danger has been recorded in the young of estuarine crocodiles and gallops over a short distance at a speed of nearly 50 km per hour have been reported (93).

Crocodiles lay oval, hard-shelled eggs which are buried in the soil or in a nest of soil and vegetation for incubation. Parental care in the form of nest guarding by the female, responding to the call of the hatching young from within the egg by opening the nest, assisting emerging young by breaking open the eggs, carrying the hatchlings to the water in their mouth, holding them in a nursery and protecting them from predators are habits common to almost all species of crocodilians. All species of crocodilians, irrespective of sex, react vigorously to the distress call of hatchlings, which is very similar in the different species.

All crocodilians are carnivorous. Three species occur in India.

MUGGER, or MARSH CROCODILE

Crocodylus palustris Lesson

LOCAL NAMES. Hindi, Gujarati, Marathi *Muggar*; Oriya *Kuji khumbhiora*; Bengali *Kuhmir*; Kannada *Mossalay*; Tamil *Muthalai*; Malayalam *Muthala, Cheengkani*; Telugu *Moseli.*

SIZE. Rarely over 4 m. Two records slightly over 5.5 m in Sri Lanka. A 3.5 m specimen shot in Madhya Pradesh weighed approximately 200 kg. Another specimen 3.75 m in length, had a girth of 1.6 m, and required twelve men to lift it off the ground.

IDENTIFICATION. In the field, the Mugger and the Estuarine Crocodile are difficult to distinguish from each other, but normally they do not occur together. Snout broad, and without distinct ridges in front of the eyes. A row of 4, distinct, sharply raised, scales just behind the head called post occipitals, back armoured with 16 or 17 transverse and 6 (rarely 4) longitudinal series of bony plates (scutes) embedded in the skin. Ventrally the skin lacks armour. Tail with two series of flattened, vertical scales merging and continued as a single row to the tip. Toes webbed.

COLORATION. Olive above with specklings of black which are more evident in young. White or yellowish white below.

HABITAT, DISTRIBUTION, and STATUS. Inhabits rivers, lakes and other large bodies of water in the plains and up to 600 m in the hills, throughout the Indian subcontinent, from Baluchistan in the west to Assam in the east, and from Nepal in the north to Tamil Nadu in the south. Also Sri Lanka and Iran. A single record from Burma. Within this range the species has now become extinct over large areas.

HABITS. The best known and most widely distributed among the three species of Indian crocodilians. All Indian river systems and their connected streams, lakes, sizeable ponds and jheels once had their quota of the mugger. In summer those living in transitory waters either aestivate or migrate during the night to more permanent sources (84). Large specimens usually spend the day basking on the bank or a rock facing the water, ready to slip in at the least alarm. The open mouth of basking crocodiles is a method of heat control. In most areas of its distribution the mugger makes burrows for habitation. In the salt lakes area of Sind (Pakistan), they were known to occupy burrows on the sides of the hills bordering the lakes. Regular trails led to the burrows which had an entrance diameter of 61 to 80 cm or more, and led through a 2.5 to 4.5 m long passage into a chamber wide enough for the animal to turn around and lie comfortably (55). Similar burrows, situated on banks of lakes, and rivers, and at the foot of trees, or below rocks by the side of streams, and other water bodies have been noted in other areas of the range of the species. The mugger

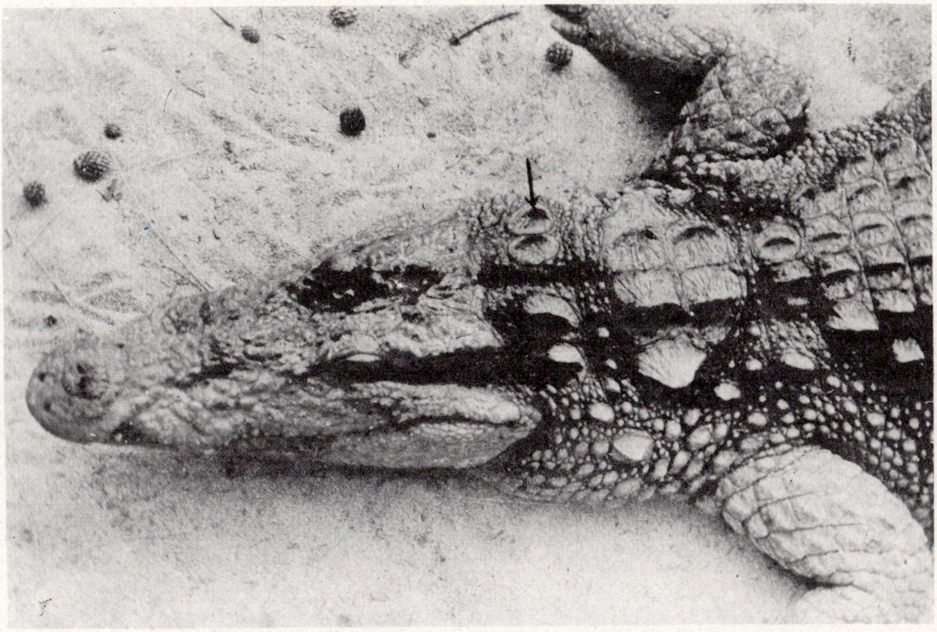

Head and neck of Mugger or Marsh Crocodile. Arrow indicates the four distinct post-occipitals behind head.

Photos: S. A. Hussain

Head and neck of Estuarine Crocodile (See pp. 10, 13)

Photo: D. K. Lahiri Choudhury
Gharial female on nest.

Photo: Robert Bustard
Gharial egg hatching (See p. 15)

is an excellent swimmer, the tail being the exclusive propellent. On land, it rests on its belly, but walks and runs with the body well off the ground. The senses of sight, hearing and smell are well developed and the animal remains very alert while basking on land.

FOOD. Hunts more or less exclusively in water; the diet is largely fish but any animal that can be overcome is taken. Once a prey has been sighted in the shallows or at the edge of the water the attack is made under water and at speed. A large prey, if several crocodiles are present, may be torn to bits and completely eaten, otherwise it is said to be stored till decomposition makes dismembering easy. A captive specimen buried in the soil the meat and fish given to it as food (5). The food varies with size, the younger animals apparently feeding largely on water insects, snails, frogs, and any fish they can catch. Crocodiles on Salsette Island, Bombay are said to feed on the fruit of the fig tree *Ficus glomerata* (1). Larger crocodiles may attack, drag in, and drown, any animal that is within their capacity to kill. The recorded stomach contents of the mugger include leopard, wild dog, hyaena, chital, sambar and nilgai fawn, four-horned antelope, barking deer, monkeys, domestic dog, goats, calves, pig, ducks and a variety of wild birds. It has also been noted scavenging on the kill of a tiger near a forest pool (26). Man-eating by this species is uncommon in India. The coarse jewellery and odd portions of the human body that are occasionally found in the stomach of killed crocodiles can usually be attributed to the habit of eating corpses. A curious habit is the swallowing of stones, allegedly as aids to digestion and as ballast in floating and diving. The size and quantity of these gastroliths, are apparently related to the size of the animal. A 3.5 m long crocodile had about a kilogramme of stones in its stomach (67).

VOICE and CALLS. Usually silent but hisses loudly when threatened facing the enemy with snapping jaws and lashing tail. Adults occasionally roar, a call said to resemble the bellow of cattle. The call is often repeated twice or thrice in quick succession (11). The significance of this call is not known. It is reported that the roar of a specimen shot within its burrow, sounded in the confined space, like the beating of a roll on a bass drum (55). From the hatchling onwards grunts form the typical mode of vocal communication between crocodiles. The distress cry of the hatchling, which is a high-pitched version of the grunt, results in prompt assistance from larger individuals in the neighbourhood which may charge the intruder or predator.

BREEDING. The scent glands are probably active during the breeding season and assist in the sexes locating each other. There is a record of a female going into voluntary captivity to mate with a male whose presence it could have known only from the effluents flowing out of the enclosure holding the male (69). Mating has been observed from mid January in the southern to March in the northern parts of the country (24, 87). Dominant males emphasize their status in the hierarchy and subordinate males show submissive gestures as for instance, raising the head to show the underside of the jaw, when near dominant males. Mating occurs usually in water, the male lying half above the female and with the tail below and cloacae opposed. Copulation lasts for up to 10 minutes. The female lays her eggs in a pitcher shaped hole about 50 cm deep and 30 cm in width dug by her in sand, earth or gravel on a stream, river, lake or tank bank, 2 to 500 m away from water. The covered nest, may be level with the ground or may show a slight mound (84). The number of eggs laid is related to the size of the female, the larger and older animals laying a larger number of eggs. Eggs 3 to 40 or more in a clutch, are white in colour, hard shelled, and range in size from 70 × 50 to 80 × 50 mm. Incubation period slightly in excess of two months but may go up to 90 days depending on prevailing day temperatures. Temperature level of the egg chamber remains constant at about 29-30°C and does not normally show a variation over 1°C. Wetting of nest by brooding female by evacuating water or urine from her cloaca, has been recorded in wild and captive specimens and probably helps in humidity and temperature regulation. The female protects the nest by lying either in the water close to the nest or on the nest itself and chasing away intruders and egg predators such as monitor lizards, jackals, and wild pigs. A captive specimen did not feed during this period. The degree of protection offered by the female varies with individual temperament.

In captivity a growth rate of 6 cm per month has been recorded when the right diet, sufficient cover and sunlight were available. Captive specimens have reached an average length of 130 cm, girth of 50 cm and weight of 8 kg at the end of the second year (86).

MISCELLANEOUS. Vitality is retained by the muscles a long time after death. A crocodile shot through the head in the morning jumped and quivered till late afternoon and the flayed body had so much movement that vultures kept away (67).

It is popularly believed that the number of stones in the stomach equal the animal's age and that 108 eggs are laid equalling in number the beads in the Hindu rosary. In Hindu mythology the crocodile is the vehicle of Niridhi, the regent of the SW point of the compass. Crocodiles are often kept in semi-captivity, usually in association with a religious establishment. Notable among these is the Mugger Pir at Karachi in Pakistan (59).

Commercial hunting and trapping for collecting the belly skin of the animal for the leather industry has now made this and other species of Indian crocodilians endangered species. However, captive breeding efforts have had very promising results. It is reported that removal of marsh crocodiles from lakes and rivers encourages the growth of coarse fish, for example the cat fishes, at the expense of commercially valuable species such as the carps.

ESTUARINE or SALT-WATER CROCODILE

Crocodylus porosus Schneider

LOCAL NAMES. Same as for the Marsh Crocodile in most Indian languages. Oriya *Baula kumbhira.*

SIZE: Probably the largest of present day reptiles. The largest skull available measures 1 metre in length and it is believed to have belonged to a specimen of about 7 metres in total length (22). Specimens over 5 m in length have been obtained in the Sunderbans and in Orissa river estuaries but are now exceedingly rare. A 4.5 m long captive specimen weighed 408 kg.

IDENTIFICATION. As mentioned under the marsh crocodile the two species appear almost identical in the field. In India, however, the Estuarine crocodile, as the name implies, is restricted in its distribution to the tidal estuaries and lower reaches of the larger rivers and hardly ever occurs in fresh water.

The snout is longer than in *palustris.* Dorsal armour of 6 to 8 longitudinal series of scutes. A strong ridge in front of eye, nearly half the length of the snout. No enlarged post occipital scales. Other characters as in *palustris.*

COLORATION. Dark olive or brownish olive above interspersed with yellow which is distinctive of this species, sometimes with a brassy tinge. Young with black markings above.

HABITAT, DISTRIBUTION, and STATUS. Inhabits tidal estuaries of the larger continental rivers, marine swamps and coastal brackish water lakes. In India the Estuarine crocodile once had a distribution extending from Vembanad lake in Kerala on the west coast through the estuaries on the east coast (Cauvery river, Tamilnadu; Brahmani river, Orissa) to the Sunderbans in West Bengal. Also in the Andaman and Nicobar islands. The species is seriously endangered, from hunting in earlier days and now largely from loss of habitat, particularly breeding sites. Recent surveys in the Andamans have brought to light the precarious position of the animal in the Andamans and Nicobars (85, 19). It is now extinct in Kerala and Tamilnadu. A sanctuary for the species has been established in the Bhitarkanika Island and adjacent areas in the Brahmani Baitarni river estuary in Orissa.

Extralimital. Sri Lanka, Burma, Thailand, Indonesia to North Australia, Philippines and the Solomon and Fiji Islands (possibly a stray).

HABITS. Generally resemble those of the mugger but it is more aquatic and has been even sighted far out at sea. In the Bhitarkanika Island area of Orissa some of the smaller animals apparently spend considerable time on land inside the dense estuarine forest. Several well-marked trails lead from the water into the forest and the sound of boats moving along the numerous creeks often flush them out from their day retreats.

FOOD. Mainly the fish occurring abundantly in its estuarine habitat. This massive crocodile also preys on fairly large animals and used to be responsible for most of the attacks on man in SE. Asia. An authentic report states that a crocodile about 5 m in length, shot in Orissa, had in its stomach the well-preserved body of a man, one leg and the head and arms severed from the body.

BREEDING. Mating commences in the dry season and egg laying with the beginning of the southwest monsoon in May on the mainland; earlier in the Andamans. Unlike the mugger, the estuarine crocodile is a mound nest builder. Nests are composed of mounds of vegetation and mud raked together by the female using her hindlimbs and tail. There is no particular preference and available vegetation is used. The proportion of mud in the composition of the mounds is related to the availability of vegetable matter. Nests are large averaging 75 cm in height and 2 m in diameter. The central spherical egg chamber of the nest houses the clutch of eggs ranging in number from 20 to 72 (average 50 eggs). The clutch size may be as low as 2,3, or 10 also. Egg size 72 × 47 to 81 × 54 mm ranging in weight from 91 to 137 g. The temperature within the egg chamber of the nest remains lower than the outside temperature during the day and higher at night. Mean temperature of egg chamber is 31.4°C (82). Incubation period varies from 80 to 90 days. The female prepares one to four guard wallows, near the nest. The depth of the wallow is dependent on soil conditions. The reaction of the guarding female is dependent on individual temperament. She may attack or escape into the water from an intruder. In a pair kept in captivity, the female lay on top of the nest and the male reacted actively against human intrusion (50). The presence of the female near the nest does not guarantee the safety of the nest which is often pilfered by natural egg predators such as the water monitor, which stealthily opens the nest from the side opposite to the wallow. The young when ready to hatch respond to loud noise or movement near the nest by grunting. The grunting hatchlings are released by the attending female opening the nest and the young are led or carried in the mouth to the nearest water. The grunting call of the young helps them to remain together and such nursery clusters are looked after by the mother crocodile for nearly ten weeks. Hatchlings measure 25 to 30 cm in total length at emergence.

GHARIAL, or LONG-SNOUTED CROCODILE

Gavialis gangeticus (Gmelin)

LOCAL NAMES. Hindi *Gharial*; Bengali *Mecho kumhir*; Oriya *Thantia kumhira*: male *Ghadiala*, female *Thantiana*; Bihari *Nakar, Bahsoolia nakar*; Nepali *Chimpta gohi, Lamthora gohi.*

SIZE: Maximum reported length 6.75 m. Believed to attain a length of up to 8 m.

IDENTIFICATION: Easily distinguished from other crocodiles by the long and narrow snout which ends in a bulbous tip. The jaws have on each side, 27 to 29 undifferentiated teeth in the upper jaw, and 25 to 26 teeth in the lower jaw. The first three teeth of the lower jaw fit into notches on the upper jaw. Adult male with a large pot-like cartilaginous mass on the tip of the snout, hence the name Gharial (*ghara* = pot).

COLORATION. Adult dark olive or brownish olive. White or yellowish white below. Young greyish brown with five irregular transverse bands on the body and nine on the tail.

DISTRIBUTION, HABITAT, and STATUS. Confined to the Indus, Ganges, Brahmaputra and the Mahanadi river systems in the Indian subcontinent and the Irrawaddy and Arakan river systems in Burma. Once very common, increasing human use of rivers has restricted the gharial to the few remaining wild stretches of its former habitat. Presently the main habitat is the Chambal, Girwa, Rapti and Narayani rivers of the Ganges system. The species is now rare and endangered.

HABITS. River dwelling crocodilians inhabiting deep pools at river junctions, and bends, called 'kunds' and the deep gorges in hilly country. They are believed to spread out with the flood waters of the monsoon and to return to the perenial 'kunds' at the end of the rains. Midstream islands and sandbanks are used for basking, particularly during the winter months (December to February) and gharials are often seen near river banks during the high water monsoon months. A facile swimmer they are clumsy on land propelling themselves with their legs in a sliding movement when coming out to bask. They rarely move far from the river bank. Larger animals have particular basking spots. Young Gharial make a groaning noise when disturbed. Adults have been heard bellowing and to make a groaning noise when in distress. Breeding males hiss loudly, the *ghara* possibly acting as a resonator.

FOOD. Predominantly fish. Occasionally takes turtles, birds and small mammals and is said to feed on corpses. The fish caught is manipulated by the jaws and swallowed head first.

Gharial stomachs contain stones as do those of other crocodiles. A 5 m long specimen had 4.5 kg of stones in its stomach, the largest weighing *c.* 230 g (29).

Stones from gharial stomachs are said to retain an offensive odour for a long time (25).

BREEDING. The mating behaviour of the gharial in its basic pattern is similar to that of other crocodilians. Mating occurs in the cold weather months of December-January. It is stated that the male uses the *Ghara* as a hook on the female's snout for leverage when mounting. Mating is in water. Females commence breeding when they are about 2.5 m in total length and males at 3 m when their age is *c.*13 to 14 years. Gharials nest in late March, early April and the nesting season is said not to vary by more than 10 days in any year and all females in an area nest more or less within a week. Nests normally sited in sand which assures high incubation success. Nest depth varies from 30 to 37 cm with a width of 22 cm. Average clutch size 40 (range 10 to 96). Eggs white, hard shelled, size 55 × 57 mm. Optimum incubation temperature lies between 32 and 34°C and the incubation period ranges from 72 to 92 (mean 84.5) days (70, 15). The young, as in other crocodilians, grunt when ready to hatch. Gharials like other crocodilians show parental care in the form of nest protection, release of young, and guarding of hatchling clusters. Hatchlings measure on an average 325-375 mm at birth, with a weight range of 75-97 g. Predation on Gharials is largely in the egg and hatchling stages from nest predators such as rats, pigs, jackals, and monitor lizards. Hatchlings are taken by birds of prey, and large wading birds and in water by large turtles and fish. Gharials have lived in captivity for 29 years.

TURTLES and TORTOISES

The reptile order Chelonia includes the marine turtles, the freshwater tortoises or terrapins, the freshwater turtles and the land tortoises. Turtles and tortoises are easily recognised from all other animals by their characteristic bony shell. The shell has two parts, the carapace above and the plastron below, joined along the flanks between the fore and hindlimbs. The shell has an outer layer of horny shields and an inner layer of bony plates, both regularly arranged. In one family of freshwater turtles and one species of marine turtles, however, skin, instead of horny shields, covers the bony shell. The body within the shell is rigid with the ribs fused to the bony plates, but the parts outside the shell, namely the neck, limbs, and tail are free-moving. Land and freshwater forms have varying ability to retract these into the shell. The evolution of the shell is obscure and fossil Chelonians closely resemble the presently living forms. Food habits vary, some species being carnivorous, some vegetarian, and others omnivorous. The jaws lack teeth but in a few species the horny beak that serves in their place has tooth-like serrations. The sexes are not easily distinguished but the male generally has a longer tail and a concave plastron. All chelonians bury their eggs in soil for incubation. The eggs may be hard or soft-shelled depending on the species. Clutch size varies from under ten in some land forms to over a hundred in marine turtles. The Chelonians are long-lived. The accepted authentic record for longevity is 152 years for a giant tortoise (*Geochelone gigantea*) which lived from 1766 to 1918 in one of the Seychelles Islands in the Indian Ocean.

MARINE TURTLES

Marine turtles lead a completely aquatic existence and, excepting the female when egg laying, normally do not come ashore once they reach the sea from the sands under which they hatched. All have paddle shaped limbs. The longer front limbs are used for swimming, being moved through water in a manner comparable to the wing-beats of a bird. The head and limbs cannot be retracted into the shell. In food habits they may be herbivorous, carnivorous or omnivorous. Turtles are known to man from the habit of coming ashore to bury their eggs under the sands of the seashore, a habit recorded as early as the 4th century A.D. in Tamil literature. Marine turtles are mainly tropical in distribution but some species occasionally enter temperate seas. Five species are known from Indian waters. The Loggerhead *Caretta caretta* is rare.

GREEN TURTLE
Chelonia mydas (Linnaeus)

LOCAL NAMES. Hindi *Samudra kacchua, Dudh kacchua*, *Yadi-da* Andamans); *Kap-troji* (Car Nicobar); *Kap-ka* (Nicobar); Malayalam *Kadalama;* Tamil *Peramai, Kadalamai*; Bengali *Kachchhap*; Gujarati *Kachabo, Durya-ni-kachbi*; Oriya *Samudra kaichha.*

SIZE. Adults may attain a carapace length of over a metre (maximum recorded 139.79 cm) and weights of up to 155 kg (60 A).

IDENTIFICATION. The carapace has four costal shields and 25 marginal shields. The carapace shields do not overlap. Limbs normally with a single claw each. A single pair of prefrontal shields on head. Jaws not hooked.

COLORATION: Adults olive, green, brown above with spots or blotches or streaks of brown or black. The female is usually more richly pigmented. The hatchling is dark blue-black.

HABITAT, DISTRIBUTION, and STATUS. Mainly confined to the warm tropical waters of both hemispheres. Adults occur in the vicinity of marine algal growth, their main food. Along the Indian coast the species is known from the shallow seas off the Gujarat coast, the Gulf of Mannar and off the islands of Lakshadweep and the Andaman and Nicobar groups. The Green Turtle is the most heavily exploited among marine turtles, nearly one million eggs used to be collected annually from the nesting beaches of the Sarawak area in Borneo alone.

HABITS. Very little information is available on the habits of the turtle apart from its breeding behaviour. The movements of the turtles after egg laying are obscure, but marked turtles have been recovered up to 1400 miles away from the beach where they were born (60 A).

FOOD. Herbivorous when adult. Feeds on marine algae and sea grass of genus *Cymodacea* etc. The preferred food plant in the Gulf of Mannar is the alga *Gracillaria* sp. (47). It is estimated that an adult green turtle requires about 4.5 kg of plant material per day.

BREEDING. The breeding habits of the Asian race of the Green Turtle (*Chelonia mydas japonica* Schweigger) are well documented from observations on those breeding on the islands off the coast of Malaya and Sarawak where turtle-egg collection is an established industry (36). The green turtle prefers to lay well above highwater mark, on island beaches with light sand and lee situations. Nesting females are most circumspect and nesting activities are confined to the hours of darkness. Movement on land resembles the breast stroke of swimmers, the body being lunged forward by action of the forelimbs. The track of a green turtle on sand varies in width from 91 to 112 cm (12). The nest digging is set in motion by the feel of soft sand under the body. Initially a 'body pit' is dug with the front flippers, lowering the animal below the beach level. The

Ridley Turtle; Green Turtle; Leathery Turtle *Del.* A. Gnanasekar (See pp. 20, 18, 22)

Photo: I. D. Kehimkar

Indian Pond Terrapin (*Melanochelys trijuga*)

Photo: S. R. Sane

Spotted Black Terrapin (*Geoclemys hamiltoni*) (See pp. 23, 26)

nest hole is then dug using the hind flippers alternately to scoop out the sand till even with the body tilted back the flippers are unable to scoop out further sand. The nest hole, oval in shape, and approximately 50 cm below the beach surface, receives a clutch averaging 104 eggs, laid in about a quarter of an hour. Eggs average 40 mm in diameter and 34 g weight. The Green Turtle is insensitive to disturbances while actually laying. On completing the clutch the hole is filled, compacted and the movements of the departing turtle effectively disguise the location of the nest. The temperature of the nest remains remarkably constant and rises to about 35°C near hatching time from the heat generated by the developing eggs. The young emerge 45 days after laying during the warmer months and 70 days of laying during the colder months. The young of a clutch hatch more or less simultaneously or in batches which work up to the surface in a group, erupting out of the sand after dark. Emergence is inhibited by sand temperatures above 28°C a built-in protection against exposure to daytime predators (57). However, the young may emerge during the day if rain lowers the sand temperature. Once above the surface, the baby turtles hasten to reach the sea, orientating to the lighter skyline of the sea. Loss through predation is enormous at this stage, the young having to face predation from a host of crabs, birds, and other predators on the shore and sharks and other predatory fishes offshore.

Green Turtles nest throughout the year but there may be seasonal peaks, believed to be May to September in the Indian region. Marked females nested six to seven times at intervals of ten days and a maximum of eleven successful nests by a single female in a season has been reported. Green Turtles have a well-marked breeding cycle and return to the same breeding beach after an interval of three years (36). Mating occurs offshore of the nesting beaches. The records of this species breeding on the mainland coast of India are only from the Gujarat coast. It is believed that the Asian race of the Green Turtle attains maturity between four and six years of age.

Olive Ridley Turtle

Lepidochelys olivacea (Eschscholte)

LOCAL NAMES. As for the Green and other sea Turtles. *Gadha kacchua* (Hindi, Andamans).

SIZE. Attains a carapace length of about a metre.

IDENTIFICATION. Distinguished by the presence of five or more costal shields on carapace. Marginal shields on carapace 27, rarely 25. Carapace of the young with three distinct keels.

COLORATION. Adult olive-brown above, yellowish below.

HABITAT, DISTRIBUTION, and STATUS. The Ridley is widely distributed in the tropics of Indo-Pacific and the East Atlantic, and is the commonest turtle along the Indian coasts. There is however, little information about its habits and behaviour.

FOOD. Omnivorous. In captivity feeds on dead fish, crabs, and other crustacea, and soft parts of molluscs.

BREEDING. The turtle most commonly nesting on the Indian shores, the Ridley has no particular breeding season but breeds throughout the year. There are possibly peak periods related to the climatic conditions. Copulating pairs have been seen in the sea off Palmyra Point in Orissa in January. Mating takes place off-shore of the nesting beaches, and the male, like in other marine turtles, does not normally come ashore. The nesting behaviour differs slightly from that of the Green Turtle. A body pit is not dug and the turtle, after filling the nest-hole, smooths the sand by turning and tamping with the body. A curious habit of the Ridley is the covering of the nest with nearby vegetation, a habit first noted and recorded by a Tamil poetess of the 4th Century A.D. She identified the plant as 'Udumbu Kodi', the local name of the Ground Glory (*Ipomea biloba*) a very common creeper on sandy beaches (64). This habit and the use of the same plant has been reported for the Ridley nesting on Krusadi Island near Rameswaram in Tamil Nadu. Other plants used at Krusadi are Ravana's Moustache (*Spinifex squarrosus*) and the herb *Lannea pinnatifida* (47). Normaly a nocturnal nester, but instances are known of the females coming ashore well before sunset to lay. The tracks on sand vary in width from 77 to 88 cm (12). The major nesting beach in India is the Gahirmata of the Orissa coast in Balasore District where over 100,000 turtles come ashore to nest in about 10 days. These large scale nesting gatherings are called 'arribhadas', a term originally used to describe such massive gatherings at several points on the Mexican Coast (60A). Large numbers used to be caught and marketed in Calcutta. The clutch varies from 40 to 125 (egg diameter 36 to 39 mm, weight 34 to 38 g). The young hatch in 54 to 56 days and take a further 4 to 6 days to emerge from the sand.

HAWKSBILL TURTLE

Eretmochelys imbricata (Linnaeus)

LOCAL NAMES. As for other marine turtles *Kangha kacchua* (Hindi, Andamans); *Tau-da* (Andamanese), *Kap sah* (Car Nicobar), *Kap kael* (Nicobar).

SIZE. Carapace length up to 850 mm.

IDENTIFICATION. Distinguished by the strongly overlapping shields of the carapace. Aged specimens may have these juxtaposed but can be distinguished from the Green Turtle (with which this species shares the common character of four costal shields) by the presence of *two prefrontal shields* on the head. Head small and narrow, with strongly hooked jaws. Limbs with two claws each.

COLORATION. Adult marbled yellow and dark brown above, yellow on the underside. Young are brown and blackish below.

HABITAT, DISTRIBUTION, and STATUS. The Hawksbill is widely distributed in tropical seas.

HABITS. The habits of the species in Indian waters are little known.

FOOD. Omnivorous, but inclined to be largely carnivorous, feeding on sponges and other invertebrates and on fish as well.

BREEDING. The Hawksbill nests on the shores of a number of islands in the Andaman and Nicobar group. The clutches usually of over 100 eggs are laid in nests from 30 to 50 cm deep. Nesting occurs throughout the year with apparently a peak during the monsoon months. The eggs vary in size from 33 to 40 mm diameter.

MISCELLANEOUS. The flesh is said to be poisonous for man at certain seasons and instances of death from eating turtle flesh are attributed to the flesh of this species. The Hawksbill is the species from which the 'Tortoise shell' of commerce is obtained. These are the scutes of the carapace removed immediately after death. In Sri Lanka they used to be removed from the live animal under the belief that the scutes are regenerated. A small cottage industry exists in the Nicobars in the manufacture of ear and finger rings, bangles, combs etc. (12).

Leathery Turtle

Dermochelys coriacea (Linnaeus)

LOCAL NAMES. As for other marine turtles *Sher kacchua* (Hindi, Andaman Islands), *Kap-chyoot* (Car Nicobar), *Kap-heebu* (Central Nicobar).

SIZE. The largest of the present-day chelonians. Adult rarely exceed 5 feet (1.5 m) in carapace length, and weight of *c.* 400 kg.

IDENTIFICATION. Shell consists of strong cartilage with an underlay of a mosaic of small bones, the whole structure is covered with smooth skin instead of horny shields. Seven longitudinal ridges above and five below. Keels well marked in the young.

COLORATION. Slate black above spotted with white, paler below. Young black with the keels defined in yellow or white.

HABITAT, DISTRIBUTION, and STATUS: The Leathery Turtle is widely distributed in tropical seas but is everywhere uncommon. Occasionally recorded in temperate waters. Along the Indian coasts the species was not uncommon at the turn of the present century. About forty used to be caught annually at Quilon, Kerala, while laying or from the sea using special nets (16). The species is now rare, along mainland coasts but not uncommon in the Andamans and Nicobars.

HABITS. The Leathery Turtle, is more oceanic than other species of marine turtles and as in other species there is very little information available on its movements.

FOOD. Leatherbacks live mainly on jelly fish and are believed to be deep sea feeders.

BREEDING. The Leathery Turtle nests on the west coast of India approximately below 11°N. latitude and in the Andamans and Nicobars. The breeding season appears to be the monsoon months and nesting has been recorded on the Kerala coast in July. Normally the female ventures out of the sea at night but there is a record of a laying in the late afternoon (41). The tracks size varies from 172 to 250 cm in width (12). The nest hole is about 70 cm in depth. Clutch size varies from 90 to 190. Eggs average 52 mm in diameter and hatch in about 70 days. The main breeding grounds of the species in this part of the world are Sri Lanka, and the east coast of Malaya.

Photo: Edward Moll

Indian Sawback (*Kachuga tecta*)

Photo: J. Vijaya

Brahminy Terrapin (*Hardella thurgi*) (See pp. 24, 25)

Kachuga dhongoka; Kachuga kachuga; Batagur baska
After Gray's *Illustrations of Indian Zoology* (See pp. 25, 26)

FRESHWATER TORTOISES OR TERRAPINS

The Terrapins or Freshwater Tortoises are hard-shelled forms which closely resemble land tortoises in appearance. All have axillary and inguinal scent glands. Herbivorous species predominate and omnivorous forms are unusual. Terrapins lay small clutches of hard-shelled eggs which are buried in the soil for incubation. The most numerous and the least known among Indian chelonians the freshwater tortoises are grouped under a single family, the Emydidae. The majority of the species occur in the large river systems of the Indo-Gangetic plain. Many are semi-terrestrial.

INDIAN POND TERRAPIN

Melanochelys trijuga (Schweigger)

LOCAL NAMES. Hindi *Talao kachuwa*; Marathi *Kasau*; Tamil *Neer amai.*

SIZE. Reported to reach a shell length of 220 mm. More information is needed on the size attained by this species and its races.

IDENTIFICATION. Distinguished from land tortoises by the flattened limbs and fully or almost fully webbed digits. Shell not very convex, with three keels, one median and two lateral. The shell margin is curved inwards laterally and behind and flares out above the hindlimbs. Snout short, skin on the back of the head smooth or divided into large shields. Limbs with enlarged scales. Tail short.

COLORATION. Shell light brown in young, darker brown or blackish in adult. Plastron has a yellow border which is prominent in the young. Four geographical races occur in India and are distinguished by the colour of the head. These are:

M. t. trijuga (Schweigger). Head greyish or olivaceous with yellow or pink reticulations. Distribution: Maharashtra, Karnataka and Tamil Nadu;

M. t. coronata (Anderson). Head olivaceous. Snout and top of head black. Temporal region yellow. Distribution: Kerala;

M. t. thermalis (Lesson). Head black, spotted or reticulated with orange or red spots. Distribution: Sri Lanka, Ramanathapuram District and Point Calimere, Thanjavur District of Tamil Nadu;

M. t. indopeninsularis (Annandale). Head grey or brown with indistinct yellowish reticulations. Distribution: Bihar and West Bengal.

HABITAT, DISTRIBUTION, and STATUS. The commonest and most widespread of Indian Terrapins. Prefers slow-flowing or sedentary waters. The species has not been recorded north of 20°N. latitude in peninsular India but it is likely to occur further north.

HABITS. Young are more thoroughly aquatic than the adults, which are semi-terrestrial. Terrapins are often seen basking on rocks and logs from which they slip into the water at the least sign of danger. Several may live together in a convenient hideout under rocks or a hole in the bank, or submerged tree trunks etc in the water. They swim in a scrambling manner, each limb working independently. The scent glands emit a strong and very disagreeable odour when the animal is disturbed.

FOOD. Forage on land at night. The food is largely vegetarian but they are known to feed on animal droppings also.

BREEDING. The breeding habits are scarcely known. The season appears to be the late monsoon months. Elliptic eggs, white in colour, and hard-shelled, are laid in clutches of 3 to 8 in holes dug by the female on land well above the water line. Several clutches may be laid in a season and hatch approximately two months after laying. The eggs measure *c.* 44 × 25 mm.

Several species of terrapins occur in the river systems of North India and countries to the east of the Indian subcontinent. The more widely distributed among these as well as some of the uncommon and rare forms are described below:

INDIAN SAWBACK or ROOFED TERRAPIN *Kachuga tecta* (Gray). Small sized (230 mm carapace length) terrapin of the Indus, Ganges and Brahmaputra river systems. The single midback keel ends in a backward pointing spine on the pentagonal 3rd vertebral shield. Carapace olive-green with small black spots with a orange or red band on first three vertebral shields. Plastron yellow in adult, orange with distinct blackspots in young. Head black above, yellow on the sides, neck black with thin yellow lines. Herbivorous.

DECCAN SAWBACK TERRAPIN *Kachuga tentoria* (Gray). Differs from the Indian Sawback in having the second vertebral shield shorter than the third. Carapace paler, head olive without red temporal patches but with a red spot behind tympanum. Occurs in the Mahanadi, Godavari, and Krishna river systems. Also in Bangladesh. A race (*K. t. circumdata*) which has been described from the Calcutta area has a pink band along the junction of the costal and marginal scutes.

CHAPANT or SMITH'S TERRAPIN *Kachuga smithi* (Gray). Known from the lower reaches of the Ganges in W. Bengal. It is more common in the Indus system in Pakistan. More or less similar in size to *Kachuga tecta* but the keel of the 3rd vertebral ends in a rounded projection. Carapace flatter, olive-brown to pale brown, plastron black (17A). Eggs in clutches of 5 to 8 are buried in sand for incubation. Carnivorous but takes vegetable food also.
Size. Carapace length 230 mm.

DHOOR OR DHONGOKA TERRAPIN *Kachuga dhongoka* (Gray). Not uncommon, medium sized terrapin of north India (Corbett NP eastward), Nepal. Distinguished by the second vertebral shield being pointed and entering the third shield. The carapace is rather rough textured, olive or brown above with a black stripe along midback. Breeding males with red longitudinal stripes on neck, red or yellow spots on throat, head red above with blue sides. Ventrally yellowish, clutches of 30 to 35, elongated oval eggs, 5.5 × 3.3 cm in size buried in sandbanks, have been obtained in March and specimens collected in that month laid after capture. Young have been obtained in June/July. Size 400 mm length. The meat of the species can be eaten even by Brahmins. However the use of its meat was prohibited in one of Asoka's inscriptions.

The SAIL TERRAPIN, *Kachuga kachuga* (Gray), the largest among this group of Terrapins, occurs in the eastern section of the Gangetic system in U.P., Bihar and W. Bengal in India and in Nepal and Bangladesh. The central keel on the back is prominent on the 2nd and 3rd vertebral shields. Shell olive or brown above yellow below. The pale brown neck has seven red or reddish brown lines. Head blue laterally. Throat with a pair of oblong red or yellow spots. The male is said to be bright red on top of head. The female lacks the red lines on the throat.

The adult is said to weigh about 20 to 30 kg and is easily distinguished from *dhongoka* by the smooth oily back and large size. Breeds in March, laying eggs on sandbanks of the Ganges. Size, shell length 390 mm (Smith). Digits fully webbed.

KHASI HILLS TERRAPIN, *Kachuga sylhetensis* (Jerdon), is uncommon and so far known only from the hills of Meghalaya and Nagaland. The carapace is domed and the median keel is in the form of a projecting spike at the posterior margin of the 3rd shield. Marginal shields 26 in number; hind margin strongly serrated. Digits fully webbed. Shell olive-brown above, keel paler, yellow below, each shield with a large brown spot. Head and legs brown, a yellow wavy stripe along back of head, another along lower jaw. Neck with longitudinal streaks. Size, 180 mm shell length. Habits unknown.

BRAHMINY TERRAPIN OR KALI KAUNTHA, *Hardella thurgii* (Gray), Widely distributed and common in Indus, Ganges and Brahmaputra river systems. Carapace somewhat flattened. The median keel disjointed and in the form of a projecting knob at the hind end of each vertebral shield. Fourth vertebral shield not longer than its width. Shell dark brown with keel and first three costals black. Yellow below with a large black patch on each shield. Head brown with four orange-yellow bands, a curved one on top of snout, another below nostril as far as the eyes, a third passing from behind eye to the tympanum and a fourth along the lower jaw. Limbs brown with yellow margins. Digits fully webbed. Tail short. A completely aquatic species said to frequent slow flowing and stagnant water. Herbivorous. The species is commercially exploited. Breeding habits little known.

Size. Shell length 500 mm. The male is said to be much smaller with 170 mm shell length. Adult female weighs about 20 kg.

SPOTTED BLACK TERRAPIN *Geoclemys hamiltoni* (Gray). An elegant species of northern India, the carapace has three well-defined keels. Plastron nearly as long as carapace, deeply notched at the back. Colour jet black above spotted and streaked with yellow. Soft parts brown or black, spotted with yellow, largest spots on head and neck. Carnivorous. Said to feed on snails in nature. Habits little known. Size, shell length 310 mm.

EASTERN HILL TERRAPIN *Melanochelys tricarinata* (Blyth). Distinguished from the Common Pond Terrapin by the yellow plastron. Little information is available on this almost terrestrial hill terrapin.

KERALA FOREST TERRAPIN *Heosemys silvatica* (Henderson). A small terrestrial terrapin with tricarinate carapace and prominent median keel. Colour black above, plastron yellow with two spots. Front of head and jaws bright yellow, a red spot on top of snout. Back of head and neck brown. Limbs and tail black. The species was recently rediscovered (77b). It had not been collected since its original discovery in 1911 when two specimens were obtained from burrows inside dense forest in the Cochin forests of Kerala. The specimens lived in captivity for six months (35A). The species is omnivorous. Breeds in the winter months. A captive specimen laid two eggs in December (77c). Related species occur in south-east Asia.

Size. 119 mm shell length.

BATAGUR TERRAPIN *Batagur baska* (Gray). One of the largest of the terrapins (shell length 590 mm), the Batagur is easily distinguished from other terrapins by the presence of only *four* instead of five claws on the forelimb. Once abundant the species used to nest colonially between January/March on sandbanks. The season's clutch of up to 50 large sized (75 mm) eggs hatch after 70 days.

The distribution of this species is peripheral to our region as it is known only from the large rivers and their estuaries in the Sundarbans in the Indian subcontinent.

Ganges Softshell (*Trionyx gangeticus*)
Juvenile

Photo: Edward Moll

Adult (See p. 28)

Photo: Edward Moll
Chitra Turtle (*Chitra indica*)

Photo: M. Krishnan
Indian Mud Turtle or Flapshell (*Lissemys punctata*) (See pp. 28, 27)

FRESHWATER OR MUD TURTLES

Freshwater turtles have a flattened disc-like shell covered with soft skin. The limbs are semi-circular, paddle-like, and have three claws. Nostrils tubular. Tympanum hidden. The head and neck are completely retractile. The neck is very flexible and extensile. Largely carnivorous in food habits. All are thoroughly aquatic and the majority of the Indian species are found in the larger river systems of the Indo-Gangetic Plain. Only two species occur in peninsular India below 20°N. latitude. Indian freshwater turtles belong to a single family, Trionychidae.

INDIAN MUD OR FLAP-SHELL TURTLE

Lissemys punctata Lacépède

LOCAL NAMES. Oriya *Panka kaichha.*

SIZE. The female is larger, reaching a carapace length of up to 275 mm. Male approximately 100 mm less in carapace length. Maximum weight recorded 4.5 kg (Pakistan).

IDENTIFICATION. Distinguished from all other species of freshwater turtles of the Indian subcontinent by the presence of skin flaps on the plastron for hiding the hindlimbs and the tail. Carapace more or less oval and covered with smooth skin which may have a plaited appearance in the young. The front and back margins of the shell can be bent down to completely hide the retracted head and limbs. Head oval terminating in tubular nostrils, Eyes lateral and rather prominent. Digits fully webbed. Tail short. Plastron with seven callosities.

COLORATION. Carapace olive-brown. Plastron yellowish or white. Two races are recognized, on the basis of colour.

INDO-GANGETIC MUD TURTLE or FLAP-SHELL *L. punctata* (*punctata*). Distinguished by the large yellow spots on the head and carapace. Head grey. The yellow sometimes in the form of a reticulation.

DISTRIBUTION. Indo-Gangetic plain and eastwards to Akyab in Burma.

PENINSULAR MUD TURTLE or FLAP-SHELL *L. punctata granosa* (Schoepff.). Distinguished by the absence of the yellow spots and the presence of three black streaks on the greenish head.

DISTRIBUTION. The Peninsula, south of the Ganges. Sri Lanka.

HABITAT, DISTRIBUTION, and STATUS. This widely distributed turtle inhabits ponds and tanks connected with or independent of the larger river systems throughout the Indian subcontinent. One of the Indian races extends into Sri Lanka.

HABITS. The commonest of the Indian Mud Turtles, existing in a wide latitude of climatic conditions. Usually an inhabitant of shallow waters and may be seen basking among aquatic vegetation. Mud turtles swim well but prefer to lie half buried in the mud and snap at passing prey with an extremely fast thrust of the long neck. The adults and young make long overland journeys during the rainy season, which is possibly the reason for the species being so widespread. Those living in non-perennial waters aestivate in the mud or move into shrub cover if available. The turtle is able to withstand prolonged starvation. A captive specimen lived for two years without taking food. Some specimens have filamentous algae growing on them.

FOOD. Omnivorous. A voracious feeder on water plants and small animals such as frogs, fishes, crustacea, and snails. Can be very destructive in fish hatcheries.

BREEDING. Mating behaviour not known. The eggs are presumably laid and hatched during the rainy season. Clutches of up to 10 eggs are buried in the soil and hatch in about nine months in artificial conditions (77 A). Eggs spherical, hard-shelled, and measure 25 to 33 mm in diameter. Hatchlings of the peninsular race have a carapace measuring about 43 × 33 mm. Carapace brown with scattered black spots. Plastron reddish yellow with a pair of brown patches. Growth rapid when the animal is young.

The rivers of the Indo-Gangetic plain are inhabited by some of the largest existing freshwater turtles.

The CHITRA TURTLE *Chitra indica* (Gray) is probably the largest among these. It is distinguished by the long narrow head with eyes situated close to the comparatively short proboscis. Head black in colour with lighter coloured streaks. An inverted chevron on neck in front of the disc. Disc olive or grey, black-spotted in the young, marked with yellow in the adult. Carnivorous in food habits. Attains a disc length exceeding 800 mm. The hatchlings are extraordinarily small with a carapace size of even *c.* 2.9 × 3.3 cm (17A).

DISTRIBUTION. Indus, Ganges river systems, Thailand, and Malaya.

The GANGES SOFTSHELL (*Trionyx gangeticus* Cuvier) is distinguished by the greenish black-streaked head and olive-green disc with black reticulation or yellow vermiculation. The eggs are said to be perfectly spherical *c.* 2.3 cm in diameter. A carnivorous species attracted to rotting flesh. DISTRIBUTION. Indus, Ganges, and Mahanadi river systems (6C). A large number are caught and sold for food. Size. Up to 700 mm in carpace length.

The DECCAN SOFTSHELL *Trionyx leithi* (Gray) has a carapace or disc coloured olive-green with lighter vermiculations. Head greenish. Black longitudinal streaks from between eyes to nape with side streaks often present; entire or broken up. The young have four eye-spots or ocelli on the disc. Carnivorous, but like other softshells omnivorous in captivity. DISTRIBUTION. Rivers of peninsular India.

SIZE. *c* 500 mm in disc length.

The PEACOCK SOFTSHELL *Trionyx hurum* Gray, of the lower reaches of the Ganges and Brahmaputra is also provided, when young, with well-marked ocelli on a olive-green disc with black reticulations. Head marbled with dark green or black lines and yellow spots. With age the ocelli and yellow spots tend to disappear.

MISCELLANEOUS: Mud turtles are often kept by religious establishments and are fed by devotees. The Ganges Softshell is kept in the bathing tank of a small temple dedicated to Vishnu at Puri in Orissa; the BURMESE SOFTSHELL (*Trionyx formosus* Gray) in a pool associated with the pagoda at Mandalay, and the CHITTAGONG SOFTSHELL (*Trionyx nigricans* Anderson) in a large tank attached to the mausoleum of a Muslim saint, Sultan Bayazid of Bastan (Iran), situated about 8 km from Chittagong in Bangladesh. This species is now believed not to exist in the free state in the wild.

Mud turtles are an essential ingredient of vedic sacrifices. A deep depression is made in the centre of the area of the *Vedi* (=Altar) and a live turtle kept in it with sufficient food to keep it alive. If at the end of the sacrifice and destruction of the altar, the turtle is alive, the sacrifice is considered auspicious. The Chitra Turtle (*Chitra indica*) is apparently the model for the "tortoise" of Indian iconography, easily recognized as it is from all other Indian forms by the peculiar shape of its head and the nearness of the eyes to the tip of the snout. Its presence in sculpture from Madras, where it does not occur is probably due to its form being established as the model for all "tortoise" avatars (6 A). The status and biology of freshwater turtles and terrapins are poorly known.

LAND TORTOISES

All land tortoises are grouped under a single family, the Testudinidae. The shell is heavy and covered with horny shields. The head and neck are completely retractile. The hindlimbs are columnar and club shaped. Tympanum distinct. Jaws with alveolar ridges. Tail short. Eggs hard-shelled, more or less spherical. Land tortoises are largely herbivorous in their food habits, and are widely distributed in the warmer parts of the world. The larger species are now restricted in their distribution to oceanic islands. Fossils of a giant species, *Colossochelys atlas,* with a carapace length of about 2.5 m, have been recovered from the fossil beds of the Siwalik Hills. A single genus occurs in India.

STARRED TORTOISE

Geochelone elegans (Schoepff)

LOCAL NAMES. Hindi *Kachuva*; Marathi *Kasav*; Tamil *Amai*; Malayalam *Amah*; Telugu *Tabelu*.

SIZE. A carapace length of 280 mm has been recorded for a female. Males are much smaller.

IDENTIFICATION. The domed carapace has conspicuous humps. Each hump has a yellow areola and radiating yellow streaks. Shell elongate, markedly convex, with the back and front margins slightly turned up. Head with small irregular shields. Hindlimbs cylinderical, club-shaped. Front limbs flattened. Large scales on limbs. Tail ends in a spur-like scale. Carapace of young hemispherical and not humped.

COLORATION. The colour pattern is distinctive. Each shield of the vertebral and costal rows of the carapace has a yellow central areola on a black background with yellow streaks radiating out from the areola giving a star-like appearance. The plastron has black rays on a yellow background. The pattern may become indistinct with age. Head and limbs brownish yellow with darker markings.

The young have the head and limbs ivory coloured with the black and yellow patterns on the shell bright and contrasty.

HABITAT, DISTRIBUTION, and STATUS. Fairly common in sandy tracts such as the coastal scrublands of the Ramanathapuram District in Tamil Nadu and the semi-arid and desert tracts of peninsular India. It is said to be very common in Chittoor district, A. P. and the adjacent Kolar district, Karnataka. Its status in wetter areas is not clear. Occurs in the peninsular area of India westward to Sind in Pakistan. Precise information is required on the status of the Starred Tortoise within this range. Not an uncommon species.

HABITS. Largely crepuscular, spending the rest of the time under cover. In the northern parts of its range it remains dormant during the cold season. The

Photo: I. D. Kehimkar
Starred Tortoise (*Geochelone elegans*).

Photo: Edward Moll
East Asian Tortoise (*Geochelone elongata*) (See pp. 30, 31)

Photo: S. R. Sane
Travancore Tortoise (*Geochelone travancorica*)

Photo: Edward Moll
Eastern Hill Tortoise (*Geochelone emys*) (See p. 31)

season of most activity is the monsoon when tortoises may be seen wandering around even during the day.

FOOD. Omnivorous but inclined to be vegetarian. Feeds on succulents like *Cissus quadrangularis,* fallen fruits, grass and similar vegetation. Captive specimens thrive on all types of soft vegetables. They are also known to feed on snails, and animal and bird excreta, and in captivity are reported to have fed on carrion.

BREEDING. Mates during the rainy season when the animals are most active. The concave plastron of the male facilitates mounting. Instances are recorded of the males shoving and trying to overturn each other. The eggs, in clutches varying in number from 3 to 7, are laid in a pit dug by the female with her hind feet, the soil being moistened with her urine. When completed and tamped down, the egg hole area becomes indistinguishable from the surrounding soil. Clutches have been laid in March, April, June, October and November. A female may lay more than one clutch in a season. The hard-shelled eggs are ellipsoidal, white, matt-surfaced, and range in size from 40 × 35 to 51 × 37 mm. Incubation may take from 47 to 147 days for eggs of the same clutch.

It takes about twenty-four hours for the young to surface after hatching. The yolk sac is external at birth but is usually absorbed in forty-eight hours. The hatchling is provided with an egg tooth for breaking the shell. Within a week of birth the young start eating greedily and on the same food as the adult. Maturity is attained in about two years.

Three more species of land tortoises are known from India: EAST ASIAN TORTOISE *Geochelone elongata* (Blyth) replaces the Starred Tortoise in eastern India and South-East Asia. Distinguished by the yellow carapace with black blotches. Nuchal shield present. It has also been reported from Nepal and Chota Nagpur in Bihar. SIZE. Carapace length up to 275 mm.

TRAVANCORE TORTOISE *Geochelone travancorica* (Boulenger). Distinguished from *G. elongata* only by the absence of a nuchal or neck shield on the carapace. Restricted to the Western Ghats of Kerala and Karnataka (Coorg) said to be common in the hills between 100-300 m. The species is an example of the Malaysian element in the fauna of the Western Ghats.

SIZE. Carapace length 290 mm.

EASTERN HILL TORTOISE *Geochelone emys* (Schlegel & Müller). The largest of the Asian land tortoises, it is distinguished by the presence of two supracaudals or tail shields on the carapace. Shell dark brown or blackish, paler in the young. DISTRIBUTION. Hills of east India and South-East Asia.

SIZE. Carapace length up to 470 mm

The AFGHAN TORTOISE *Testudo horsfieldi* occurs in the Baluchistan province of Pakistan. It is easily distinguished from other tortoises in the Indian subcontinent by the presence of only four instead of five claws in the forelimb and the almost circular carapace.

LIZARDS

A widely diversified group of reptiles, the lizards are generally distinguished from the snakes by the presence of limbs. However limbless lizards occur but the Indian species can be separated from snakes by the presence of eyelids. The shape of the body is indicative of the habit of the species. Ground dwelling forms and those which live among rocks are flattened dorso-ventrally; arboreal forms are compressed laterally and subterranean forms are elongated and cylindrical. Indian lizards, except varanids, rarely exceed 500 mm in total length. The limbs show adaptations related to their mode of life and in a few burrowing and sand-dwelling forms they are reduced or absent. In arboreal forms the eyesight is good, the eye may be reduced in size in burrowing forms and in many species which live under the soil the lower eyelid is provided with a transparent window permitting vision with the eyelids closed. The majority of lizards have well-developed ears capable of hearing sound carried through the air or through ground vibrations. The sense of smell is not uniformly well developed. The tongue is distinctive for each family and is a useful guide for their separation. The skin has scales, at least on a portion of the body. Lizards are mainly insectivorous, a few are herbivorous and some are carnivorous. Except for a few ovo-viviparous forms Indian lizards lay small clutches of eggs with calcareous, parchment-like, or hard shells. The eggs are usually buried in the soil for incubation. Lizards occupy all terrestrial biotopes in India from the deserts to evergreen forests and from plains level to 5000 m in the Himalayas. Seven families, Gekkonidae (Geckos), Agamidae (Agamids), Chamaeleonidae (Chameleons), Scincidae (Skinks), Lacertidae (Lacertids), Anguidae (Glass Lizards), and Varanidae (Monitors) occur on the mainland; an additional family, the Dibamidae, occurs in Nicobars. None of the Indian lizards are poisonous.

GECKOS

A distinctive group of lizards with world-wide distribution in different types of habitats approximately between 40° latitude north and south of the Equator. In size they vary from 40 to over 350 mm in total length. The body is flattened in Indian geckos and the skin is thin and granular on the back in the majority of forms. The eye lacks eyelids and is covered by a transparent shield in all species except the Fat-tailed Geckos of the genus *Eublepharis*. In some classifications this genus is separated as the family Eublepharidae. The digits of many arboreal geckos are provided with lamellae (plates) on the toes to assist in climbing. In Indian geckos this feature shows a maximum development in the genus *Hemidactylus*. These have on the surface of the toe-lamellae a mat of setae (hair-like structures) each ending in a pair of spoon-like cups less than a micron in width. Several theories have been advanced to explain the ability of geckos to climb smooth vertical surfaces and to move on ceilings defying gravity. Early theories suggested adhesion by secretions under digits but there are no glands on the digits. Subsequently the minute setae under the toes were believed to act as suction cups, a theory that was given a boost by the examination of the structure of the setae through a scanning electron microscope. Another theory advocated frictional forces. However none of these considered the part played by the foot as a whole in gecko movement.

This problem has been examined in a careful anatomical study of the foot bones and muscles of the large Tucktoo Gecko (63). Apparently several systems are involved. A key bone in the hind foot acts as a swivel permitting the foot to hold firmly while the whole body is pivoted on it. Unlike animals which walk on the soles of the feet (plantigrade) where the toe is the last part of the foot to leave the ground, in the geckos the tip of the toe leaves the ground first and the toes are the last to touch the ground when the foot is put down again. The muscles that control the precision movement of the foot show several peculiarities. Unlike other running lizards and birds where the limbs are thin and only the tendons activate the limbs, the legs and feet of geckos are heavily muscled. The fibres of the muscles that control the toe movements are arranged like the barbs on a feather from a central tendon permitting more muscular attachment to the setae holding plates. Between the toe bones and the setal plates blood sinuses occur and when the gecko flexes its toes the pressure on the sinuses helps to press the setae hard against the climbing surface. The exact nature of the action of the setae is still not clear. The 'suction cup' theory is not tenable because a dead gecko can also be made to hold on to a vertical surface and because the setae lose their holding ability if the animal is repeatedly pulled off from the wall. The delicate nature of the setae, the large number of spoon-shaped tips and the ability of the animal to bring them into close contact with the substratum suggest that adhesion is a 'dry adhesion' brought about by surface phenomena possibly at the molecular level between the setae and the substratum.

Most geckos have a voice. The Fat-tailed Gecko, however, produces a mechanical sound by rubbing specialised areas of the body. Geckos may be arboreal, ground, and rock dwellers. Some are commensal with man. One species, occurring in the Nicobar Islands in the Indian territory, has the ability to glide through the air. Indian geckos are nocturnal spending the day in a convenient hideout, an exception being the Andaman Emerald Gecko (*Phelsuma andamanensis*) which is diurnal. The majority of Indian species are soberly coloured in shades of brown and grey. The colour of the adult and young may show remarkable variation. The main food is insects but the larger species eat any animal that they can overcome including other geckos and the smaller snakes. The geckos are in turn preyed upon by a large variety of animals.

Geckos obtain a certain amount of protection from the ease with which the tail breaks off and moves actively after detachment, thereby drawing attention to itself and away from the animal. The tail is beautifully adapted for this purpose. The individual vertebrae of the tail have a central unossified plane along which the tail breaks. Other adaptions apart from the cleavage plane of the tail bones, is the segmentation of the scales, fat layer, and the muscles by non-cellular matrix and the presence of a sphincter in the caudal artery before each breaking plane which automatically seals the artery when the tail breaks. The writhing and wriggling of the broken tail is believed to be due to the presence and breaking of nerve fibres in the spinal cord known an Rissner's fibres which are normally responsible for the automatic control of the flexure and pose of the body. The tail regenerates but does not attain the original colour or shape.

All geckos are oviparous, two eggs forming the normal clutch except in very small species which lay a single egg. The egg shell is calcareous except in the eggs of the Fat-tailed Gecko which has a parchment-like shell. The shell is soft when laid but hardens on contact with air. A sticky coating when laid enables them to stick to each other and to the surface on which they are laid. A suitable egg-laying area may be used by several geckos.

Eggs of some species of House Geckos of the genus *Hemidactylus* have hatched after experimental exposure to sea water for periods of up to eleven days (13). This hardiness is perhaps one of the reasons for the wide distribution of the members of this genus. The incubation period varies from 40 to more than 70 days.

One species, *Hemidactylus garnoti,* occurring within Indian limits in the eastern Himalayas and eastern India, is believed to reproduce parthenogenetically, without fertilization from another individual. The absence of males considered in association with a triploid chromosome number of 70 and the non-rejection of skin grafts between different individuals leads to the conclusion that the species is unisexual (27, 44).

None of the geckos are poisonous.

Photo: I. D. Kehimkar

Northern House Gecko (*Hemidactylus flaviviridis*).

Photo: S. R. Sane

Brook's Gecko (*Hemidactylus brooki*) (See pp. 35, 37)

Termite Gecko (*Hemidactylus triedrus*).

Photos: S. R. Sane

Rock Gecko (*Hemidactylus maculatus*) (See p. 37)

NORTHERN HOUSE GECKO

Hemidactylus flaviviridis Rüppell

LOCAL NAMES. Hindi *Chhipkauli, Cheechauk*; Gujarati *Garoli*; Oriya *Jhitipiti*; Marathi *Pal*; Telugu *Balli*; Kannada *Hulli*; Tamil and Malayalam *Palli*; Bengali *Ţiktiki*.

SIZE. Maximum recorded length 180 mm of which half is the tail.

IDENTIFICATION. The genus *Hemidactylus* which includes all the common Indian house geckos can be distinguished from other geckos by the strongly dilated digits with two series of lamellae or plates on the underside and a free

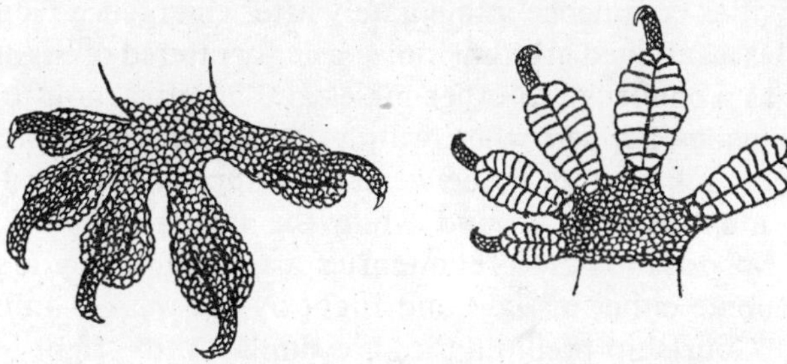

Dorsal and ventral view of the foot of *Hemidactylus*

slender-clawed terminal phalange to each toe. The skin of the back is granular with scattered or regularly arranged larger tubercles. Pupil vertical. Body flat. The different species of the genus cannot however be separated with certainty from each other unless a specimen is in the hand. The presence of 7 to 10 lamellae under the 1st toe and 11 to 14 under the 4th distinguishes *H. flaviviridis* from other *Hemidactylus*. The skin of the back has few or no tubercles.

COLORATION. Pale grey above, markings indistinct, ventrally yellowish. In the Punjab, the colour is recorded as showing a seasonal change, increasing in intensity with increase of temperature from pale grey in March to dark grey with five wavy black stripes in June (60).

HABITAT, DISTRIBUTION, and STATUS. A commensal of man, usually seen on the inside walls of buildings. Occurs throughout India north of latitude 20°N. Also in Bangladesh, Pakistan and up to the shores of the Red Sea. The commonest House Gecko of North India, rare in the Peninsula except in the cities of Bombay Deccan. Probably widely introduced through human agency.

HABITS. House Geckos resemble each other very closely in habits and the description of the habits of *flaviviridis* is generally applicable to other species of the genus *Hemidactylus*. Geckos are creatures of habit and have a strong sense of territory, occupying the same diurnal retreat day after day. A hunting territory is normally maintained. But if food is abundant, as for instance in the vicinity of an electric light, the territory may not be vigorously defended and intruders are not chased away. Usually the largest animal appropriates the best site. Activity is seasonal. The cold weather is spent in a torpid condition in a secluded hideout from which they emerge as the weather becomes warmer (late February in Bombay, March in northern India). Nocturnal; retreating into a hideout during the day. Geckos move with speed and dexterity on walls.

FOOD. Mainly insects. Smaller insects are swallowed whole; larger victims are battered to death and to a manageable softness and eaten. Dangerous insects are generally avoided. Occasionally cannibalistic.

BREEDING. Activities commence immediately after emergence from the winter torpor. The call is a subdued grinding noise usually uttered on sighting another gecko and acts as a challenge to other males and an attraction to the female. Fights between males happen when wandering males enter established territories. The fights follow a set sequence; a slow approach towards each other followed by a manoeuvring around while the tail is twitched slowly and sinuously as a cat does. These preliminaries are followed by a sudden rush which may terminate either in fight and flight by the weaker animal or flight without a fight. Courtship preliminaries are similar to the fight behaviour up to the final confrontation which in courtship is a chase and capture sequence of the female by the male. When copulating the male holds the female by the neck and partially covers her with his body. Contact is effected by the male twisting his hind end to oppose the female's cloaca. Mating occurs in March-April (65).

Eggs, usually two, rarely one or three in a clutch, are laid in April-May. The hard-shelled eggs are sufficiently sticky when laid to remain attached to each other and to the substratum. Eggs are laid in a crevice, a hole, or a secluded dark corner. In a suitable location several clutches may be laid, by different females. Eggs range in diameter between 9 and 14 mm, weigh *c.* 900 mg and are slightly oval in shape. Depending on the air temperature the incubation period varies from 33 to 54 days. Young hatch out in May-June and measure *c.* 55 mm in total length. A double egg tooth is present. Growth is rapid and the hatchlings attain adult size in 3 to 4 months and breed the year after birth (65).

Several other species of *Hemidactylus* geckos occur commonly in India. The more widely distributed species are:

BARK GECKO (*Hemidactylus leschenaulti* Dumeril & Bibron). Length 166 mm, closely resembles *H. flaviviridis* but differs in having 6 to 7 lamellae under the 4th toe. A common gecko of the Indian peninsula, Sri Lanka and Pakistan. It is the house gecko in some parts of the country (e.g. Bhubaneshwar, Calcutta). Frequents avenue trees, particularly banyan and tamarind which have

Photo: I. D. Kehimkar
Banded Rock Gecko (*Cyrtodactylus dekkanensis*).

Photo: S. R. Sane
Tuctoo (*Gecko gecko*) (See pp. 38, 39)

Fat-tailed Gecko (*Eublepharis macularius*)
Juvenile

Photos: S. R. Sane

Adult (See p. 38)

numerous nooks and crannies on the trunk. The dark grey colour of this gecko with wavy bands on the back merges with the colour of the tree bark in such situations. Breeds between April and July. *Eggs* are laid in pairs and a single female may lay several clutches. Eggs hatch after approximately 32 days.

BROOK'S GECKO (*Hemidactylus brooki* Gray) Length 135 mm. Back with conical tubercles arranged in regular rows, 5 to 6 lamellae under 1st, 7 to 10 under 4th toe. Colour brown or varying shades of grey with brown spots, whitish below. The commonest of the Indian *Hemidactylus* geckos occurring throughout the Indian subregion. Lives in a variety of habitats, on trees, rocks, under stones and on buildings. Its loud *chuck chuck chuck* call is often heard after dusk. Breeds as other geckos do during the hot weather. Two spherical eggs are laid. Eggs hatch in about 39 days. Widely distributed in Asia and Africa and has been introduced elsewhere in the tropics of the world. This species is perhaps the best example of unwitting distribution by human agency.

SOUTHERN HOUSE GECKO (*Hemidactylus frenatus* Schlegel). Length 125 mm. Throat with small granular scales, back with scattered tubercles. Lamellae 4 to 5 under 1st toe, 9 to 10 under 4th. The gecko most often seen in houses in South India. Occasionally on trees. Breeds in the hot weather. The eggs in clutches of two are laid in April-May and hatch after 42 days. The call resembles that of Brook's Gecko and is equally loud and this is perhaps the noisiest of Indian geckos. Widely distributed in South India, Sri Lanka, SE. Asia and through human agency to North Australia.

TERMITE HILL GECKO (*Hemidactylus triedrus* Daudin). Length 170 mm. A handsomely marked gecko with three white-edged olive-green crossbars on back; greenish above eye. Back with large tubercles arranged in 16 or 18 rows. A ground-dwelling form commonly associated with termite hills. Widely distributed in India and Pakistan. Breeding habits not recorded in India. In Pakistan females with eggs have been collected in May-June and young are said to be plentiful in September-October (56).

ROCK GECKO (*Hemidactylus maculatus* Dum. & Bibr.). Length *c.* 270 mm. Perhaps the largest of the Indian *Hemidactylus* geckos. Usually associated with rock formations and rock-cut caves. Occasionally on the rough outer walls of buildings and on trees. Easily distinguished by its large size and distinctive coloration consisting of dark spots on a grey background. The bands and spots are prominent in the young. Lamellae 9 to 10 under the 1st and 11 to 13 under the 4th toe. Breeds in the summer. *Eggs* two in a clutch, and 19 × 16 mm in size. Feeds mainly on insects but is known to take other geckos. Occurs in the Peninsula from the Surat Dangs of Gujarat southwards.

TUCKTOO (*Gecko gecko* Linn.). Length *c.* 340 mm. The common house gecko of southeast Asia, found in India only in Bihar, Bengal and Assam. The largest among the geckos. The common name is derived from its loud call *tuck too* which in quiet surroundings carries for over 100 m. A clutch of two eggs (average length 25 mm) is laid in July and hatch 64 days later (88). Hatchling 90 mm in length and brightly coloured being reddish brown above with white spots. Below greyish brown. Tail banded black and white.

GLIDING GECKO (*Ptychozoon kuhli* Stejneger). Length *c.* 190 mm. A remarkable gecko which occurs within the Indian region in the Nicobar Islands. Distinguished by the webbed toes and the skin-flaps on the sides of the head, body, and tail; the latter is frilled. Grey or brown above with darker markings. The cryptic colour in conjunction with the skin-flaps makes the animal indistinguishable when on the bark of a tree. Eggs, in clutches of two, hemispherical in shape, and when stuck to the bark of a tree, resemble mushrooms and are thus easily overlooked. *Eggs* are laid in November-December and hatch after 10 weeks. The skin-flaps act as parachutes for the rigidly held body and tail, helping the animal to glide facing into the wind at an angle of 52° to 53° (74). A gliding distance of *c.* 13 m was covered by an animal dropped from a height of 10.5 m (76).

FAT-TAILED GECKO (*Eublepharis macularius* Blyth). Length *c.* 210+mm. Distinguished from all other Indian geckos by the *presence* of eyelids. The Fat-tailed gecko is of robust build with a distinctive swollen tail. The young are banded with yellow and white crossbands on a brown background. Adults spotted or reticulated with yellow or bluish grey. Inhabits arid country. A nocturnal gecko, it lives largely on insects and other arthropods but also preys on other geckos. Produces a squeaky mechanical noise by sudden sideways jerks of the head. Breeds in the hot weather. Two eggs with parchment-like shell are laid, size 31 × 13 mm. In India occurs in Gujarat, Rajasthan, and as far south as Pune in the Deccan. A specimen lived in captivity for 11 years.

Several other genera of geckos, some widespread, others restricted in distribution occur in the Indian Region.

BANDED GECKO (*Cyrtodactylus* spp.). Size of species ranges from 80 to 192 mm. Easily distinguished from other geckos by the *vertical pupil* and the *absence of dilatations* on the toes. Most species are conspicuously and handsomely marked in life with spots or bands on a grey or brown background. Nocturnal. Usually dwell among rocks and on the forest floor. Some are semi-arboreal. A widely distributed genus occupying a variety of habitats in the western and eastern Himalayas, the deserts of Kutch, and the forests of Western Ghats, Eastern Ghats and the Andamans. Twelve species are known.

Among these, *Cyrtodactylus k. kacchensis* (Stoliczka) of Kutch, *C. rubidus* (Blyth) of the Andamans, *C. stoliczkai* (Steindachner) and the closely related *C. lawderanus* (Stoliczka) of the Western Himalayas and *C. dekkanensis*

Photo: Gertrud Newmann & Helmut Denzau
Common Garden Lizard (*Calotes versicolor*) male.

Photo: I. D. Kehimkar
Forest Calotes (*Calotes rouxi*) male in breeding colour (See pp. 42, 44)

Photo: *Zahida Whitaker*
Southern Green Calotes (Calotes calotes).

Photo: A. R. Rahmanı
Fan-throated Lizard (*Sitana ponticeriana*) male in partial display
(See pp. 44, 45)

Cyrtodactylus kacchensis (After Smith)

(Günther) of the Western Ghats are not uncommon. Like other geckos they feed largely on insects.

DWARF GECKO (*Cnemaspis* spp.). Species range in size from 64 to 117 mm in total length. Distinguished by the slender *undilated digits and round pupil.* All are brown above, spotted or marbled with darker colour. In some species a light coloured vertebral band is present. Ground geckos, are partial to rocks in wet, hill forest country. One species *C. littoralis* (Jerdon) is arboreal and diurnal. Eleven species are known from India all restricted to the forests of the Western Ghats with the majority of species occurring in the southern part of the Peninsula. One species *C. kandiana* (Kelaart) has become a commensal of man and is widely distributed (Andamans, Malay Archipelago).

GOLDEN GECKO (*Calodactylodes aureus* Beddome). Size 185 mm. The terminal phalange of the slender digits *with two plate-like expansions* (one in the inner digit). Pupil vertical. Reported from rocky ravines in the Eastern Ghats. The colour in life is said to be a brilliant golden yellow. Rare.

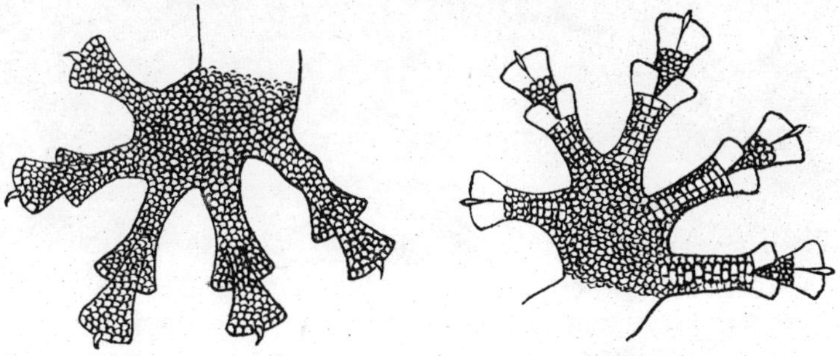

Dorsal and ventral view of foot of *Calodactylodes aureus* (After Smith)

SOUTHERN FOREST GECKO *Dravidogecko anamallensis* (Gunther). Size 95 mm. Distinguished from *Hemidactylus* geckos which it resembles by the presence of *single* lamellae beneath the dilated digits. A small nondescript grey-brown gecko widely distributed in forest areas in South India.

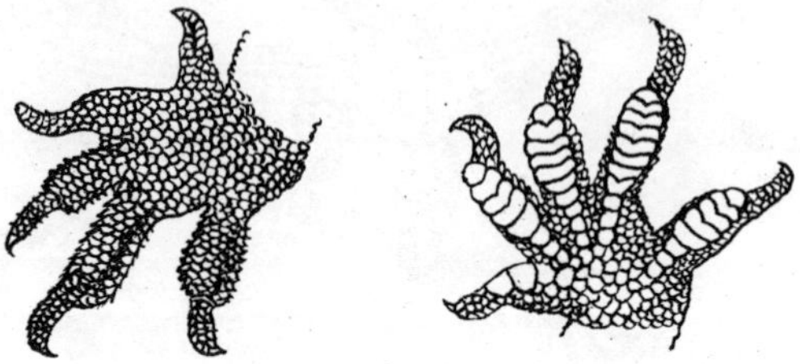

Dorsal and ventral view of the foot of *Dravidogecko anamallensis* (After Smith)

FRILLED HOUSE GECKO *Cosymbotus platyurus* (Schneider). Size 125 mm. Distinguished from other house geckos by the *fringe of skin along the sides of the body.* Grey or greyish brown, often handsomely spotted with darker colour on the back and bands on the tail. Occurs in NE. India, Nepal.

EASTERN HOUSE GECKO *Gehyra mutilata* (Wiegmann). Size 120 mm. Distinguished from the *Hemidactylus* house geckos by the absence of a free distal or last phalange on the inner digit. A pale greyish house gecko of SE. Asia and islands of the Indian and Pacific oceans. Occasionally reported from port cities (Cochin) in India.

Hemiphyllodactylus typus aurantiacus (Beddome). Size 70 mm. Distinguished by the *vestigial inner digit* from other geckos with dilated toes having ventral lamellae. This race of a widely distributed SE. Asian species occurs as a ground gecko in the hill forests of South India (type locality Shevroy Hills, Tamil Nadu). Brown above, spotted or marbled with darker colour.

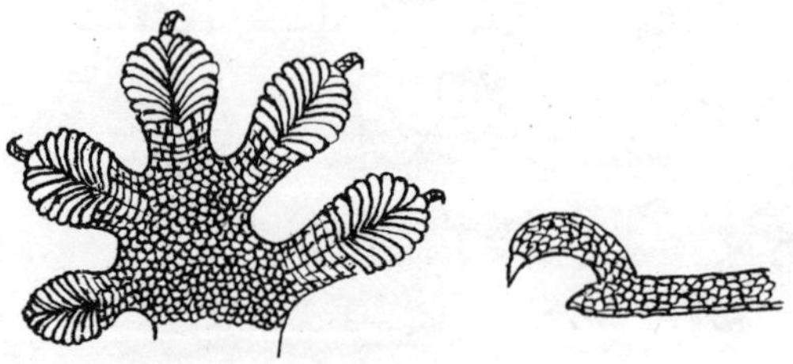

foot, ventral view and toe of *Gehyra mutilata* (After Smith

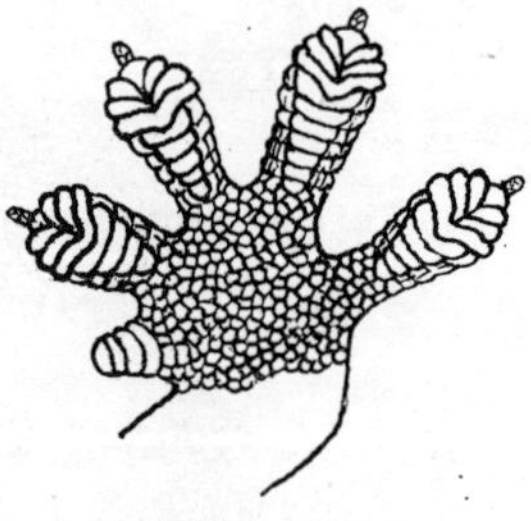

ventral view of, foot of *Hemiphyllodactylus t. aurantiacus*

SCALED GECKOS (*Teratolepis, Lophopholis*). Distinguished from all other geckos by the *imbricate or overlapping scales* on the body and from each other by the single (*Teratolepis*) and double (*Lophopholis*) lamellae under the toes. Represented by a single species each.

BANDED SCALED GECKO *Teratolepis fasciata* (Blyth). Size 67 mm. Greyish brown above with five longitudinal dark brown dorsal stripes. Six rows of white spots on back, a white band on nape. Tail with white crossbars. Reported from Sind (Pakistan), peninsular India and Meghalaya (Shillong). *Lophopholis scabriceps* (Annandale). Size 95 mm. Greyish brown with darker bands above. Reported from Tamil Nadu and Sri Lanka.

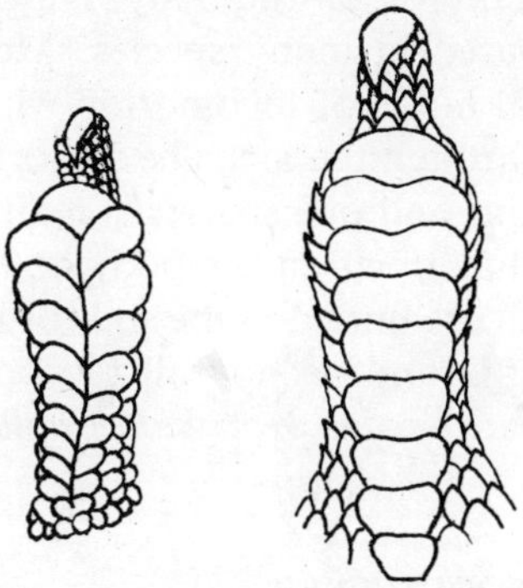

(a) (b)

underside of toe of (a) *Lophopholis sca — briceps* (b) *Teratolepis fasciata* (Enlarged) (After Smith)

Teratolepis fasciata

EMERALD GECKO *Phelsuma andamanense* Blyth. Size 136 mm. Restricted in distribution to the Andamans. A dainty little, diurnal gecko, easily distinguished by its racquet-shaped digits and brilliant coloration of rich emerald green with red spots above and bright yellow below. Tongue bright red. The species is common near human habitations. Coconut trees in flower are apparently preferred haunts because of the nectar and insects attracted by the flowers. Territories are maintained and adult males intimidate by rising on their legs to display red spots on head and base of tail. The young are believed to hatch out during the early monsoon (89). The genus *Phelsuma* has a peculiar distribution. Allied species are not from mainland Asia but occur in the oceanic islands of Mauritius, Seychelles, Reunion and Madagascar.

AGAMIDS

The Family Agamidae is confined to the Old World and the majority of genera occur in the Oriental Region. Agamids can be separated from all other Oriental lizards by the nature of their teeth. These are divided into incisors, canines and molars. In general appearance and habits agamids bear a close resemblance to the iguanas of the New World. As among other lizards the shape of the body is indicative of the habit, arboreal forms are flattened laterally and ground-dwelling forms are flattened dorso-ventrally. The types of scales on the body and their arrangement are good aids for distinguishing between species. The head which is held off the ground on a distinct neck, has small scales and lacks shields. Eye and ear well developed. Eye with eyelids. Pupil round. Spines and other appendages are common on the body. During the breeding season the male is brilliantly coloured in many species. Most agamids can move fast and some adopt a bipedal mode of locomotion when hard pressed. One genus has the ability to glide through the air. The majority of agamids are insect-eaters. A few are carnivorous and one species is mainly herbivorous. Except a few ovo-viviparous (eggs hatch within the body of the female) forms, agamids lay soft-shelled eggs which are buried in the soil for incubation. Agamids are widely distributed in India and occupy all biotopes from sea-level to over 5000 m. A species of Toad-Agama (*Phrynocephalus*) has been recorded from 17,000 ft (5182 m).

COMMON GARDEN LIZARD OR BLOODSUCKER

Calotes versicolor (Daudin)

LOCAL NAMES. Hindi *Girgit*; *Kirkantio* (Rajasthan); Gujarati *Kakido, Kachundo*; Bengali *Girgiti*; Oriya *Endua*; Tamil *Ohnan*; Malayalam *Ohnthoo*; Kannada *Vothi, Vonthi*; Telugu *Thota balli*.

SIZE. Maximum recorded for the male 490 mm (140 body, 350 tail). Females are considerably smaller. Animals from the Indian peninsula are consistently larger than those from the Indochinese region (Assam and eastwards).

IDENTIFICATION. A medium-sized, arboreal, lizard with oval head and laterally compressed body. In the male, the cheeks are muscular and swollen. Two distinct spines on each side of head behind tympanum. Dorsal scales large, equal sized, keeled and directed backwards and upwards. 35 to 52 scale rows round mid body. Tail, long, cylindrical, and swollen at the base in the male. A distinctive dorsal crest of lance-shaped scales from nape to above vent in the male.

COLORATION. Brown or sand grey above, uniform or with a pattern of spots and bars on the back and sides. Female and young usually with two lateral stripes with patterns in between. New born and young feebly iridescent golden yellow with brown patterns.

Photo: I. D. Kehimkar

Common Skink (*Mabuya carinata*).

Photo: S. R. Sane

Snake Skink (*Riopa punctata*) (See pp. 53, 55)

Photo: S. R. Sane
Jerdon's Snake-eye (*Opisops jerdoni*)—magnified.

Photo: I. D. Kehimkar
Spiny-tailed Lizard (*Uromastyx hardwickii*) (See pp. 56, 47)

HABITAT, DISTRIBUTION, and STATUS. Occupies all biotopes from dry desert to thick forests, from the plains to about 2000 m. Widely distributed throughout the Indian subcontinent and most of SE. Asia. The commonest agamid lizard of India.

HABITS. An arboreal, diurnal lizard of gardens, hedges, scrubland, and forest. Prefers shrubs and undergrowth. Adept at making itself inconspicuous by its immobility and colour. Usually sidles to the back of the bole or branch when noticed. An agile climber, it moves with speed and dexterity when necessary. Recent observations indicate that the preferred temperature for activity is 36°C. and it has a temperature tolerance ranging from 24°C. to 40°C. Temperatures above 45°C. are lethal. In northern India the lizard is inactive during the cold weather and remains in hiding, in sheltered spots.

FOOD. Primarily insects, ants forming a large proportion of the food. Small birds, nestlings, frogs, and other small animals are also occasionally taken. There is one report of the species feeding on unripe, cultivated beans (22 A).

BREEDING. Seasonal, April to end September. At the commencement of the breeding season the males become conspicuous by their pursuit of the female and by their brilliant coloration when excited. The head, shoulder, parts of the forelegs turn bright scarlet or crimson (hence the name Bloodsucker) with black patches on the sides of the throat. Each male maintains a territory and displays from an elevated site within it, usually the bare trunk of a tree from where it can see the surroundings and can itself be seen. The display, in the form of 'press ups' and a solemn nodding of the head, acts both as courtship and threat display depending on the sex of the animal it is directed at. Display by rival males often ends in brief fights with the combatants standing on their hindlegs, wrestling and biting till one of them turns and runs and is chased out of the territory by the victor. Mating is brief, the male holds the female and twists his tail under hers to copulate.

The testes are enlarged (17 × 9 mm) from June to August when they start decreasing in size in a remarkable manner and measure not more than 2.5 × 0.5 mm by January. Ovaries are indistinguishable in December but gradually increase in size and eggs are fully formed by the end of May. *Eggs*, ovoid with a soft leathery white shell, and vary considerably in size (10 × 4 to 19 × 13 mm). The variation is probably related to the water absorbtion by the developing egg. The female deposits the eggs in a hole dug by her in a flower-bed or similar soft soil areas. The hole *c.* 8 to 10 cm deep is dug in about two hours using the forelimbs alone. The clutch size varies from 11 to 23 and a clutch is laid in about half an hour with an interval of about a minute between eggs. The soil is scraped back with the forelimbs, tamped with the snout and made indistinguishable from the surroundings. Incubation period varies from 37 to 47 days depending on the temperature. The size of the eggs increases with development being most rapid between the 4th and the 18th day. Hatchlings measure 72 to 76 mm in total length and attain sexual maturity in 9 to 12 months and commence breeding the following year (7, 8).

Three other species of *Calotes* are likely to be commonly seen. The genus *Calotes* is distinguished from all other agamids by the *equal* sized dorsal scales.

FOREST CALOTES (*Calotes rouxi* Dum. & Bibr.) (length *c.* 250 mm) is widely distributed in the forests of the Peninsula especially in the Western Ghats. A comparatively smaller lizard than *versicolor,* it can be distinguished by the presence of a distinct dark fold across the neck in front of the shoulder and two slender spines of each side of the back of the head. Normally brown above with patterns and easily confused with young *versicolor.* During the breeding season the head and a narrow strip along the mid back of the male turn bright brick red contrasting vividly with the black on the rest of the body. An arboreal species. Breeds seasonally (May to September) and a clutch of 4 to 9 eggs (11 × 6 mm) is laid.

SOUTHERN GREEN CALOTES *Calotes calotes* (Linn.) is a slender, long-tailed (body 130 mm, tail 500 mm) arboreal species, widely distributed in Sri Lanka and the southern peninsula (Kerala and Tamil Nadu). The scales of the back point backwards and upwards. Bright grass-green dorsally with whitish or cream transverse stripes on the body. Prefers well-wooded country and has been collected from tall trees. A female was observed at Trivandrum laying six eggs in a 5 cm deep hole dug by her. The hole when filled and smoothened after the laying was indistinguishable from the surrounding soil.

EASTERN GREEN OR JERDON'S CALOTES (*Calotes jerdoni* Günther). Length 385 mm. Occurs in the hill areas of NE. India and is reported as common at Shillong. A bright green lizard with a pair of brown, black-edged, dorsal bands and a conspicuous fold ahead of the shoulder. Tail green at base and posteriorly with bands. Habits similar to those of other calotes lizards. Mating couples, eggs, and hatchlings have been obtained in August. Hatchlings *c.* 70 mm in total length.

Photo: Binod C. Choudhury

Chameleon (*Chamaeleon zeylanicus*) (See p. 51)

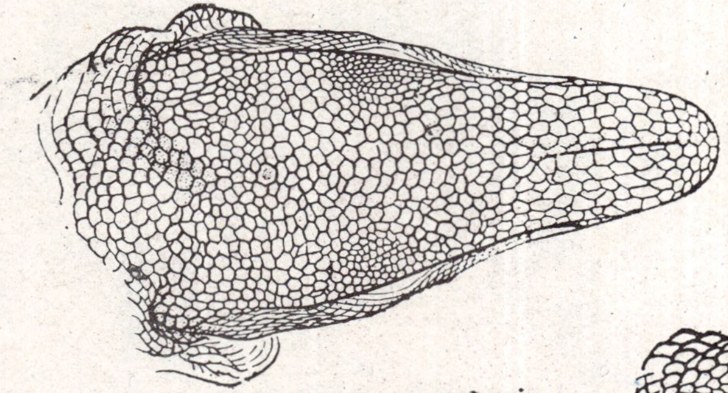

1. *Varanus bengalensis*

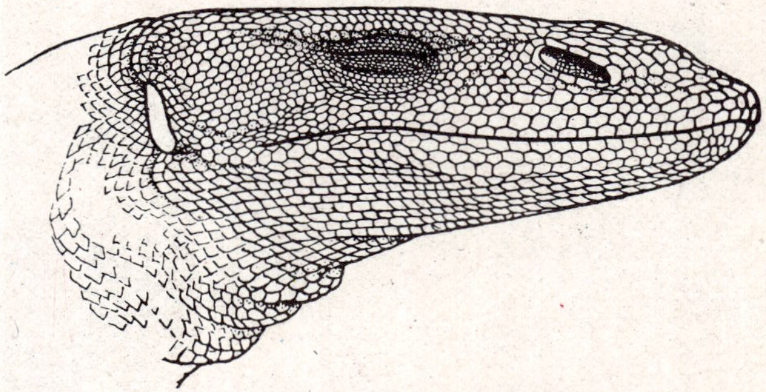

2. *Varanus bengalensis*

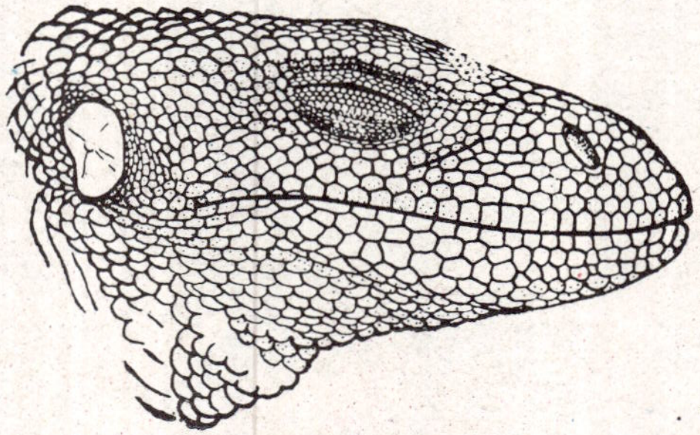

3. *Varanus flavescens.*

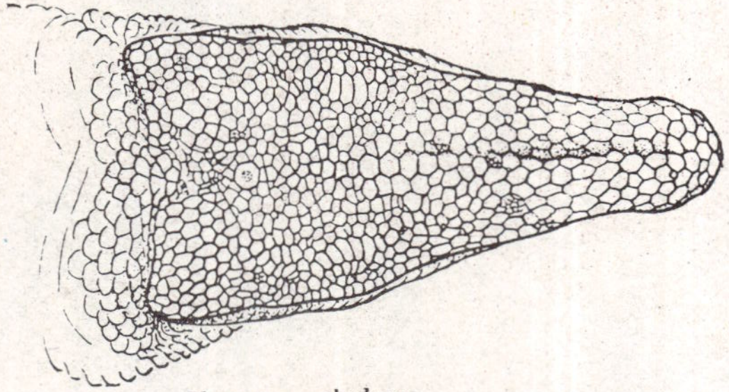

4. *Varanus nebulosus.*

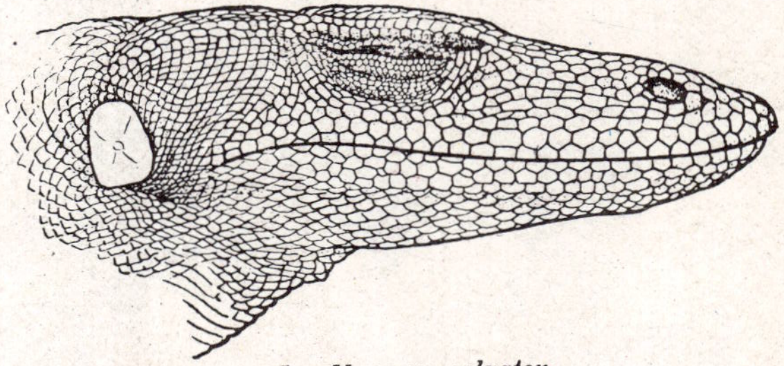

5. *Varanus salvator.*

Heads of *Varanus* (*After* M. A. Smith) (See pp. 58-60)

FAN-THROATED LIZARD

Sitana ponticeriana Cuvier

LOCAL NAMES. Marathi *Sargota*; Tamil *Veeseri ohnan*.

SIZE. The largest specimen in the Bombay Natural History Society's collections is 205 mm in length. Two forms are recognised, a larger form (70-80 mm body length) from the Bombay area and a smaller form (40-50 mm body length) from the other areas of distribution.

IDENTIFICATION. A small lizard, easily distinguished by the presence of only *four* as against five toes in all other agamids. The fan-like throat appendage in the male is an additional distinguishing character.

COLORATION. Brown above with a series of darker, black-edged, vertebral spots along the midback, flanked by a light line on each side. Whitish below. The throat fan of the male is brilliantly coloured in the breeding season, being blue anteriorly, turning blue-black in the centre and reddish posteriorly.

HABITAT, DISTRIBUTION, and STATUS. Inhabits all biotopes except perhaps the heavy rainfall forests and the deserts. The preferred habitat is scrub and sandy country. Widely distributed from Kanyakumari to the foot of the Himalayas. Not reported east of the Ganges in eastern India. A common species.

HABITS. A ground-dwelling, diurnal agamid, common in open sandy scrub country. A fast and graceful runner, occasionally adopting a bipedal gait when hard pressed. Rests in the shade of bushes and other cover and may climb these for basking.

FOOD. Ants and other small insects.

BREEDING. Seasonal; commences in April-May when the throat fan of the male assumes its brilliant coloration. The courting male displays to the female hidden near by by raising itself on its forelegs and rapidly opening and shutting its multi-coloured throat-fan.

The testes are enlarged in May and June during which months the females are also gravid. Egg-laying commences in July and individuals may lay up to October. Eggs are laid in a hole dug by the female and the clutch size varies from 11 to 14 eggs. A female under observation dug a 6 cm deep hole in 40 minutes with her forelegs and straddling the hole laid a clutch of 11 eggs in 13 minutes, arranged the eggs with her snout, replaced the soil with her forelegs and tamped down the soil with her snout. One captive female laid three clutches of 11, 13, and 14 eggs in a period of 41 days between 29 August and 10 October. She had mated before each clutch was laid. The eggs are chalky white in colour and average 10 × 6 mm (18, 77). Incubation period not known. Juvenile males can be distinguished at an early stage by the appearance of the throat-fan.

DRACO, or FLYING LIZARD

Draco dussumieri Dum. & Bibr.

LOCAL NAMES. Hindi *Udtha chipkali*; Tamil *Parakum ohnan*; Malayalam *Parakum ohnthu*.

SIZE. Maximum recorded length 230 mm. Male slightly smaller than the female.

IDENTIFICATION. The presence of the patagium or 'wing membrane' on the flanks between the limbs is distinctive. In *dussumieri* the patagium is normally supported by the lengthened last six (occasionally seven) ribs. Throat with a finger-like appendage which in the male is thrice as long as in the female and extends beyond the snout when erected.

COLORATION. Coloration cryptic, resembling the bark of the trees on which the lizards live. Ash-grey or darker above with a median series of black circles. Head with a pair of cross bands. Ventrally greenish yellow. Patagium brown above turning purplish black flecked with yellow at the margin. Ventrally yellow margined with black patches. Throat appendage bright lemon-yellow. The bright colours of the 'wing' membranes are an example of 'flash coloration', a protective device designed to draw attention to the animal when the colours are displayed in motion but making the animal disappear into the background abruptly as soon as movement ceases.

HABITAT, DISTRIBUTION, and STATUS. Inhabits primarily evergreen biotopes up to 1000 m. Also arecanut, coconut and betel vine plantations in the plains country. Restricted to southwest India from the hills near Kanyakumari to the forests of Goa. Common in many parts of Kerala but distribution erratic. The gliding lizard is an example of the Malayan element in the fauna of southwest India. All other species of the genus *Draco* occur in the eastern Himalayas and further east.

HABITS. An exclusively arboreal lizard, gliding from tree to tree with the aid of its patagia or wing membranes. The glides are usually from a higher to a lower elevation, the lizard losing height in flight. The lizard assumes a vertical position just before landing on a vertical tree trunk which it does with an audible plop. Glides covering a distance of more than 20 m have been recorded. The length of a glide obviously depends on the height from which it is launched and is possibly assisted by following winds. On landing the animal runs up the trunk in a jerky manner very like birds of the Tree-Creeper family. The lizards are most active during the morning and evening hours and rest during the mid-day heat.

FOOD. More or less exclusively ants. A stomach examined was fully packed with the red ant (*Oecophylla smaragdina* Fabr.)

BREEDING. Seasonal; courtship occurs between February and April in Kerala.

The male maintains a courting territory from which other males are excluded. On sighting a female the male bobs its head and rising on its forelegs erects and folds the throat appendage. The forwardly directed appendage is also vigorously vibrated. The body colour turns a conspicuous silver-grey and with raised body and stiff tail the male slowly circles round the female occasionally touching her rump with the extended throat appendage. If the female is receptive she moves towards him vibrating her small gular appendage, and mating occurs. Gravid females have been collected in Kerala between July and September. A captive female dug a 5 cm deep hole in the soil and laid four eggs on 25 July. The eggs, white in colour and with a partially calcified shell measured 14 × 8 mm and weighed 0.54 gm. They hatched out after 50 days on 13 September. The young at birth measured 84 mm in total length. The patagium was short and the young unlike the adult could move comfortably and speedily on the ground (39, 40).

SPINY-TAILED LIZARD
Uromastyx hardwickii Gray

LOCAL NAMES. Panjabi *Salma*; Hindi *Sanda*; Gujarati *Sandho.*

SIZE. Male average length 415 (396-489) mm; female 375 (340-396) mm. The male has a longer tail.

IDENTIFICATION. The small rounded head, with a short snout, flattened body and the tail with its whorls of spiny scales are distinctive. The skin of the body is loose and has a wrinkled appearance. Tail thick at the base depressed like the body and has cross rows of enlarged spiny scales, the spines on the sides being the largest. The spine rows are separated from each other by 4 to 6 rows of smaller scales.

COLORATION. Yellowish brown, sandy or olive with black spots and vermiculations. A distinctive black spot on the front of the thigh.

HABITAT, DISTRIBUTION, and STATUS. Inhabits the dry and desert tracts of upper India from Uttar Pradesh in the east, Rajasthan, the Kutch area of Gujarat to Pakistan. Common but patchily distributed.

HABITS. Generally found in slightly elevated areas of its habitat where it is not likely to be flooded out of its burrow. In the Banni area of Kutch it is reported to exist in large colonies on the outskirts of villages situated on the bets or islands which stay above water during the rainy season (54). Each animal lives solitarily in a tunnel excavated by itself. The sloping tunnel which has an opening of about 7 cm diameter flush with the ground may zig zag and wind for over 2 metres usually ending in a chamber where the animal can turn around (2). The lizard usually lies basking near the entrance of its burrow and is very alert, slipping into the burrow at the slightest sign of danger. Spends the four months of the winter in hibernation. Becomes active in spring and by the end of the monsoon and before hibernating for the cold weather acquires long strips of fat on each side of the backbone on which it presumably lives through the winter (61).

FOOD. Largely herbivorous. The teeth are adapted for a plant diet. The central two incisors of the upper jaw are replaced by an extension of the jawbone forming a cutting surface opposed to the fused incisors of the lower jaw. Canine absent. Feeds mainly on the flowers and fruits of the Khair (*Capparis aphylla*), the beans of *Prosopis spicigera* and the fruits of *Salvadora persica*, as well as grass. In locust-breeding areas feeds on nymphs and adults of the locust.

BREEDING. Mates in spring after emergence from hibernation. Specimens collected in Kutch in the latter half of February had enlarged (17 × 12 to 24 × 14 mm) testes and 12 to 20 developing ova (largest 5 mm) in the ovaries. The mature eggs are white shelled and of the size of a pigeon's egg.

MISCELLANEOUS. The flesh of the lizard is said to taste similar to that of chicken. The melted fat, a yellowish liquid, does not congeal in the Indian cold weather and is much esteemed as an embrocation. From its alleged medicinal properties the animal is a regular stock-in-trade of itinerant medicine men and is kept in captivity by the cruel practice of dislocating the backbone.

Other lizards of the family Agamidae are locally common in parts of the Indian region. The Agama lizards (genus *Agama*) are generally distributed in the Himalayas in the Indian region.

KASHMIR AGAMA *Agama tuberculata* Hardwicke & Gray

SIZE. Body 95, tail 150 mm. The commonest species in the western Himalayas. A medium sized lizard, bluish or greenish grey or occasionally brownish in colour. Body depressed. The male has a callous patch of scales on the belly. A diurnal lizard, it lives singly or in pairs in holes, and crevices of rocks and stone walls, scrambling over them with great agility. Feeds on ants, butterflies and other insects and is destructive to garden plants as it nips off flower petals. Males territorial and pugnacious in the breeding season. Breeds in May, June and July, possibly in the first half of August also. Small clutches of 7 to 9 eggs are laid. Hibernates in the cold weather but may be seen basking on sunny days. Abundant in the summer months in the vicinity of Simla.

SHORT-TAILED AGAMA *Agama minor* Hardwicke & Gray

SIZE. Body 90, tail 80 mm. Widely distributed in the Gangetic plains and Central and Western India. Distinguished by the tail being *shorter* than the head and body. The females are larger than the males and are more brilliantly coloured during the breeding season. It is said to be a sluggish, crepuscular and nocturnal animal. Gives out a brief squeak when caught.

Head of Agama minor (After Smith)

The SALEA lizards are peculiar to the higher altitudes (*c.* 1800 m and above) of the Western Ghats. Two species occur. The NILGIRI SALEA *Salea horsfieldi* Gray.Size. Body 95, tail 250 mm. A bright green lizard with conspicuous crest along the back extending on to the tail in the male. Common in the Nilgiris and Palni hills of Tamil Nadu. ANAMALAI SALEA *Salea anamallayana* (Beddome). Size. Body 110, tail 200 mm. Occurs in the Anamalai and Palni hills and is distinguished from *horsfieldi* by the *presence* of a strong fold in front of the shoulder, the dorsal scales being very unequal and by the nuchal and dorsal crests being continuous. Brown above with white spots on head with four V shaped marks on back. Limbs and tail banded.

(a)

(a) Nilgiri Salea

(b) Anamalai Salea (After Smith)

(b)

PENINSULAR ROCK AGAMA *Psammophilus dorsalis* (Gray). Body 136, tail 290 mm. Replaces the Himalayan agamas of the genus *Agama* in the Peninsula. A medium-sized lizard distinguished by the regularly arranged scales, the flattened body, and the presence of a fold in the skin of the throat. Predominantly a rock-dweller from the plains up to 2000 m. Common at low elevations. The coloration is cryptic particularly of the female, matching that of the rocks among which they live. It is an agile and wary diurnal lizard, darting into rock crevices at the least sign of danger. In the breeding season, the male becomes brilliant crimson red in colour on the head and foreparts of the body, and black elsewhere. The male maintains a territory and displays from a conspicuous location by 'press-ups' and nods of the head. Insectivorous in food habits. Gravid females had up to eight eggs in the ovaries. Occurs from Bihar in eastern India, and along the hills of the Eastern Ghats south to Kanyakumari. Common in the vicinity of Bangalore.

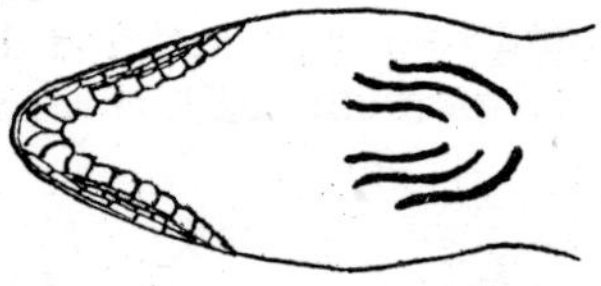

Neck of Blue-throated Lizard (After Smith)

BLUE-THROATED LIZARD *Ptyctolaemus gularis* (Peters). Size: Body 80, tail 170 mm. A species common around Shillong in Meghalaya. Recognised by the three longitudinal dark blue folds on the throat converging at the base to form a U-shaped figure. Olive-brown above with darker crossbars.

The JAPALURA lizards, *Japalura major* (Jerdon) common in the Western Himalayas and *Japalura variegata* Gray and *Japalura tricarinata* (Blyth) of the eastern Himalayas, resemble lizards of the genus *Calotes* in general appearance but are distinguished by the scales of the body being unequal in size and heterogeneous. The Japalura lizards occupy at higher altitudes the niches occupied by the *Calotes* lizards lower down. The TOAD-AGAMAS of the genus *Phyrnocephalus* inhabit deserts usually at high altitudes. One species has been recorded from Rajasthan. Small-sized lizards, the Toad-Agamas are specially adapted for life in the desert. The body is flattened. Eyes small and provided with projecting eyelids to prevent entry of sand. The nostrils can be closed. The lizards are agile and readily bury themselves in the sand when alarmed. *Phrynocephalus theobaldi* Blyth has been obtained in the Himalayas from elevations up to *c.* 5000 m. *Phrynocephalus euptilopus* Alcock & Finn has been recently recorded from the Rajasthan Desert. The Toad-Agamas are insectivorous.

CHAMELEONS

Arboreal lizards mainly Madagascar and East African in distribution. One species occurs in India. Chameleons are unique in the possession of an extensile tongue, independently moveable eyes, and in the modification of the toes into two sets of opposed clasping organs similar in appearance to the toes of parakeets. The compressed body and prehensile tail are characteristic. They possess to a remarkable degree the ability to change colour.

INDIAN CHAMELEON
Chamaeleon zeylanicus Laurenti

LOCAL NAMES. Hindi *Girgit*; Gujarati *Sarado*; Bengali *Bahuroop*; Oriya *Kuasapa, Bahurup endua*; Tamil *Pachai ohnan*; Malayalam *Pacha ohnthu*; Kannada *Hasuronthi*.

SIZE. Largest specimen measured had a total length of 375 mm.

IDENTIFICATION. A laterally compressed arboreal lizard, with a conical casque on top of the head. Body covered with granular scales. Eyes large and except for a small aperture for the pupil, covered by the granular, scaled, lid. Tympanum absent. Tongue very extensile and club-shaped at the tip. The digits of the hand in two opposed sets, two directed away from and three towards the body. The number is reversed in the arrangement in the foot. Tail prehensile.

COLORATION. The ability of the chameleon to change its colour almost instantaneously is proverbial. Basically the colour is green to which patterns of yellow and black are added in the form of bands and spots. Occasionally a uniform yellow or grey. The ability to match colours with the background is apparently limited to shades of green and yellow. The change of colour may be in response to light, heat, the emotional state of the animal and the colour of the environment.

HABITAT, DISTRIBUTION, and STATUS. The chameleon is arboreal and prefers wooded areas but is uncommon in very heavy rainfall regions. In semi-desert regions; found on hedges and shrubs. Restricted to oases in desert areas. Distributed throughout the Indian peninsula south and west of the Ganges. The western distribution extends up to Peshawar in Pakistan. Unlike the Garden Lizard or *Calotes,* the chameleon is uncommon.

HABITS. A diurnal lizard perfectly adapted for an arboreal existence. Its movements are deliberate and slow. The colour of the animal provides camouflage among the leaves. The eyes are capable of independent movement and their position on the head also permits binocular vision. When necessary the chameleon can increase its striking range by supporting itself on its hindlegs and tail and reaching out with the body. On the ground its walk is slow and stilted, each leg being waved in a curious vacillatory or hesitant action before being set on the ground. The whole effort is reminiscent of the laboured unsteady walk of a person after a protracted lay-up in sick bed.

FOOD. The feeding habit of the chameleon is unique. The tongue is inordinately long and has a viscid club-shaped tip. Captive specimens have been seen to extend the tongue as much as 30 cm beyond the mouth. In capturing prey, the eyes first focus on the prey, and once aligned the tongue is shot out and retracted with the prey sticking to the knob-like tip. The whole action is exceedingly rapid and is capable of being immediately repeated. Eight dragonflies were taken one after the other by a captive chameleon within 62 seconds. A certain minimum distance from the prey is necessary for rapid action. The food is predominantly insects. Heavy insects may drop off from the tongue because of weight and the tongue is probably ineffective against insects with highly polished bodies. An instance is recorded of a captive chameleon taking several small frogs in quick succession (75).

BREEDING. In captivity a pair mated in October. There was no courtship. The female descended to the ground in early November (9th) and dug a hole in the soil with her forelegs. The next day was spent in digging furiously and the night was spent within the hole, the animal emerging the next day after laying the clutch of 31 eggs. The hole *c.* 30 cm deep was filled up by pushing in the soil with the front feet and ramming it in with the hindlegs. The eggs were a perfect oval in shape and *c.* 12 × 5 mm in size (75). Clutch size varies from 22 to 33 and eggs average 16 mm in length. Incubation period 81 days to approximately nine months, young at birth measure 30 mm (83).

SKINKS

The skinks, Family Scincidae, are widely distributed in the tropics of the world but are comparatively more common in Australasia than elsewhere. The Americas have the least number of forms. Skinks except in the Australian Region are generally small-sized lizards, the largest species in the Indian Region does not exceed 500 mm in total length. Body covered with smooth, or keeled shiny scales, imbricately arranged. Head with distinct shields. There is typically little or no neck region. Body elongated and flattened dorso-ventrally. Limbs may be well developed, vestigial, or absent. The lower eyelid, in many burrowing or sand-dwelling forms, develops a transparent window which in some skinks completely covers the eye. In such cases the eyelids may be fused. The tongue in skinks is olfactory in function. Most skinks are ground-dwellers, the preferred habitat of the family being sandy ground. Several genera live in forests among the litter on the forest floor. Many of the forest-dwelling forms are semi-arboreal. Skinks feed on insects. The breeding habits of the majority of species are little known.

COMMON OR BRAHMINY SKINK

Mabuya carinata (Schneider)

LOCAL NAMES. Hindi *Bhabani, Loten*; Gujarati *Sáni mashi, Aroo-ni-mashi*; Marathi *Surpa chi mousi, Sapsurulli*; Kannada *Arani*; Malayalam *Arana*; Tamil *Aranai.*

SIZE. Maximum recorded length 290 mm (165 mm tail).

IDENTIFICATION. Several species of the genus *Mabuya* occur in India and are distinguished from all other skinks, except those of the genus *Eumeces,* by the presence of well-developed limbs and supranasal shields (see Fig. below) on the head. *Eumeces* species usually have *broader* dorsal than flank scales as against the *uniform* sized body scales of the *Mabuya* skinks. The common skink can be separated from all other species of the genus *Mabuya* by its scaly lower eyelid, 30 to 34 scales round the body with those on the dorsal side having 3, 5 or 7 keels. The fourth toe has 14 to 18 lamellae, 4 on the underside.

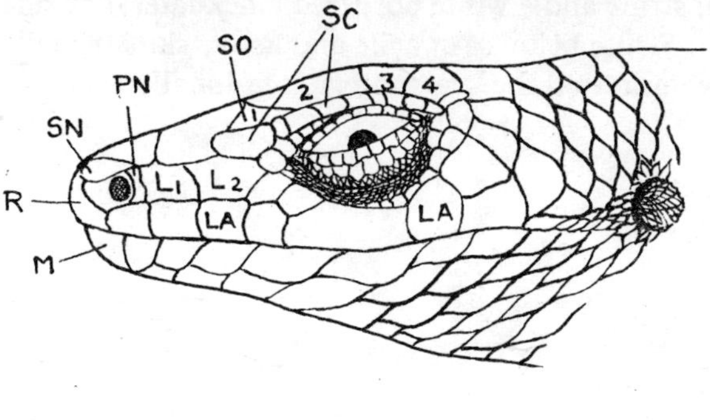

Head of *M. carinata*
R. Rostral, SN-Supra-nasal, PN-Postnasal S0,1,2,3,4, Supraoculars SC-Superciliaries, M - Mental, La-Labials L1, L2-Loreal. (After Smith)

COLORATION. Shiny brown, olive or bronze above, darker spots often present. Flanks darker. A light band from behind the eye to the base of the tail. Upper lip white. Lower parts white or yellow. Flanks of the male scarlet during the breeding season.

HABITAT, DISTRIBUTION, and STATUS. Almost a commensal of man. Equally at home in semi-urban areas and the forest. Primarily a ground-dweller. Actively searches through the ground litter for prey. Widely distributed in the Indian peninsula but rare or absent in NW. India. Common in Bengal. This is the commonest and best known of Indian skinks.

HABITS. A diurnal lizard, often entering houses in its incessant search for prey. In the forest it is more often heard than seen as it creeps through and among the litter on the ground. Occasionally 'surfacing' to survey the surroundings. The lizard returns to a particular night roost.

FOOD. More or less exclusively insects, but known to take small vertebrates also.

BREEDING. Courtship and other aspects of breeding behaviour not known. The common skink is ovo-viviparous. The uterine eggs average 15 × 10 mm. The young are seen in May and June. Up to eight young are born in one litter.

18 genera of skinks holding 109, species occur in the Indian subregion. Most of these are uncommon or rare forms. Two widely distributed species, and other species locally common in particular areas of the country are:

LITTLE SKINK *Mabuya macularia* (Blyth). A slender skink with snout to vent length not exceeding 75 mm. 28 to 30 rows of scales round the body, 5, 6, 7 or 9 keels on the scales and 12 to 17 lamellae under the fourth toe. The colour-pattern of this widely distributed species varies. The general body-colour is brown with or without spots. Oviparous. Females with eggs (5 mm diameter), three to four in number, have been collected in June. The Little Skink is common in forested areas of the Peninsula.

SAND SKINK *Mabuya bibroni* (Gray). Size, body 50, tail 65 mm. A common skink of the sand dunes of the sea coast from Orissa to Kanyakumari District of Tamil Nadu. Distinguished by the transparent 'window' in the lower eyelid. Dorsal and lateral scales with 5 to 7 keels (3 in juveniles). Brown above with a black bordered light vertebral stripe and a white bordered black lateral or side stripe from eye to base of tail. White below. An agile and active skink usually seen foraging among the vegetation of the sea shore. Occasionally reported

Window in lower eyelid of Sand Skink (After Smith)

from inland. A closely allied species *Mabuya dissimilis* (Hallowell) with 2 or 3 keels on the scales occurs in the Indo-Gangetic plains and is common in the dry western district. Light brown with three greenish white stripes bordered with black line or spots. Edge of eyelids bright yellow.

SNAKE SKINK *Riopa punctata* (Gmelin). Maximum snout to vent length recorded is 85 mm. An elongated, snake-like slender skink with feeble five-toed limbs. Lower eyelid with transparent disc. General colour brown above, each scale having a basal black spot. In the juvenile the spots are joined to form 4 to 6 lines on the back. Yellowish white below with each scale having a black central spot. The juvenile's tail is red in colour. Widely distributed but not often seen as it spends most of its life underground. Its habits are little known. Occurs throughout the Peninsula.

SANDFISH *Ophiomorus tridactylus* (Blyth). Maximum total length recorded is 185 mm. Peculiar to the sandy tracts of northwestern Rajasthan and Pakistan. Inhabits sand dunes, living just below the surface. It is nocturnal and feeds largely on beetles and termites. Hibernates in winter.

In the Himalayas small skinks of the genera *Lygosoma* (without transparent disc in lower eyelid) and *Leiolopisma* (with transparent disc in lower eyelid) are locally common.

Both genera occur in the Western Ghats where *Lygosoma dussumieri* Dum. & Bibr. is a common skink in the plains of south Kerala. The species is semi-arboreal in habit, in forested areas.

Leiolopisma himalayanum (Günther) is reported to be common in Kumaon between 4000-9000 ft in damp situations and abundant in the environs of Nainital lake and the gardens of Simla. Males with a orange or bright reddish lateral stripe in summer (Himachal Pradesh).

Leiolopisma sikkimense (Blyth) is common in the Darjeeling area, in holes in moss covered revetment walls from where they emerge to sun themselves on bright days. Lays eggs in late June. Young hatch in July.

LACERTIDS

The lizards of the family Lacertidae are poorly represented in the subcontinent. They are distinguished by the well-developed limbs, notched tongue, presence of symmetrical shields on the head, difference in the size of the scales between the dorsal and ventral sides of the body and the presence of femoral or thigh pores on some of the scales on the inside of the thigh. Two genera, *Cabrita* having *free* eyelids and *Opisops* with the eyelids *fused,* are widely distributed in the Peninsula.

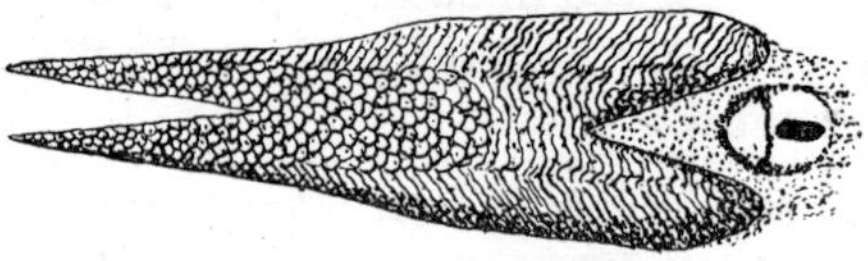

Tongue of Lacertids (enlarged) (After Smith)

LESCHENAULT'S CABRITA *Cabrita leschenaulti* (Milne-Edwards) having the anterior labial or lip shields keeled and JERDON'S CABRITA (*Cabrita jerdoni* Beddome) with labial shields lacking keels are common in the open dry jungles of the Peninsula. Both have more or less similar colour patterns, brownish or golden above with two light stripes.

JERDON'S SNAKE-EYE (*Opisops jerdoni* Blyth) is a widely distributed species which is not uncommon. Distinguished by its rough head shields which are keeled and striated. In scrub country inhabits stony areas. They are extremely fleet of foot and move so rapidly that at close quarters it is difficult for the eye to follow their movements. Breeds in May-June. The eggs are laid in June-July. Recorded clutch size 7. Size of eggs 6.5 × 4 mm.

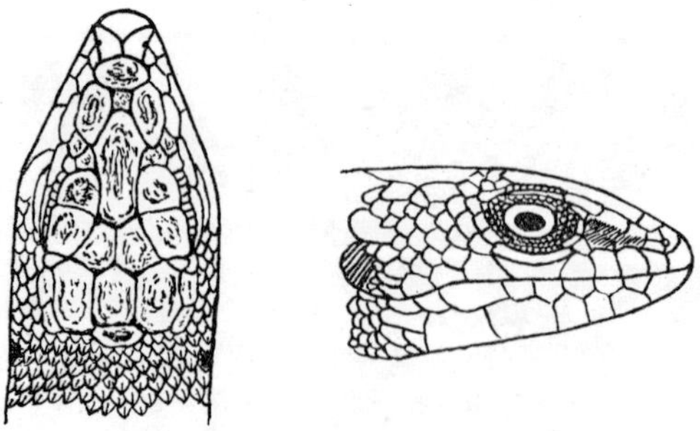

Head of Jerdon's Snake - eye (enlarged) (After Smith)

Glass Snakes

The Burmese Glass Snake *Ophisaurus gracilis* (Gray) of the family Anguidae is common at lower elevations in the hills of eastern India. There is a single record from near Simla in the Western Himalayas (59 C). The body is elongated and snake-like and lacks limbs. Glass snakes can be immediately distinguished from the true snakes by the presence of eyelids. In the Shillong area, the Burmese Glass Snake (Local Name *Naingbaen*) is not uncommon. In colour the adult is light or dark brown above with a darker lateral band having transverse series of blue, black-edged, spots. Lives under logs and stones and feeds on insects and earthworms. Breeds during the rainy season. The female stays with the eggs. Eggs collected in September in Shillong hatched in the latter half of the month. Clutch size 4 to 5 (eggs 18 × 10 to 21 × 11 mm, weight 1.5 g). The hatchlings 114 mm in length including the 70 mm tail. The colour of the hatchling is pinkish buff with a metalic sheen. A black band on each side of the body. A fine black line beneath the eye passes along the lower lip to the vent. The glass snakes hibernate during the cold weather. The tail breaks as in the geckos and has the same function of attracting the attention of the predator.

Burmese Glass Snake (After Smith)

MONITOR LIZARDS

The monitor lizards, family Varanidae, show a uniformity of structure not seen in other groups of lizards. A single genus, *Varanus,* holds all the species. Monitors are distinguished by their long and flattened body, long tail, long neck and the extremely elongated, slender, forked, tongue similar to that of the snakes. Eyes with well-developed eyelids. Teeth recurved. Head covered with small scales. Body covered with small round or oval scales. Ventral scales arranged in regular rows. Limbs well developed and the digits are armed with strong claws. Monitor lizards, often mistakenly called Iguanas, are widespread in the warmer regions of the Old World from Africa, through West Asia to Australasia in the east. The family is of ancient origin; fossil forms similar to existing species were present over 50 million years ago. The largest existing lizard is a monitor, the Komodo Dragon, *Varanus komodoensis* of Komodo Island in Indonesia, which grows to a length of over three metres.

COMMON INDIAN MONITOR

Varanus bengalensis (Schneider)

LOCAL NAMES. Hindi *Gho*; Gujarati *Patla gho*; Bengali *Goh-sáp*; Oriya *Godhi*; Marathi *Ghorpad*; Kannada, Tamil & Malayalam *Oodoombu*; Telugu *Ooder*.

SIZE. Maximum length recorded 1750 mm (tail 1000 mm).

IDENTIFICATION. Distinguished from the Desert Monitor *Varanus griseus* by the compressed tail and from the yellow monitor (*Varanus flavescens*), and the Water Monitor (*Varanus salvator*) by the nostril being nearer to the eye than to the tip of the snout. But this character is not very reliable.

COLORATION. Adult olive, grey, or brownish above with sparse black spots, yellowish below, uniform or flecked with black. Young brightly coloured, dark olive with white eye spots arranged across the back alternating with dark bars or spots. Head with lighter coloured spots. A dark streak on the temple.

HABITAT, STATUS, and DISTRIBUTION. The Common Monitor is widely distributed and lives in all biotopes from evergreen forests to the fringes of the desert. The commonest among the four species of monitor lizards which occur in the subcontinent, but in some areas, the yellow and desert monitor are more common.

HABITS. A diurnal lizard though more active in the mornings and evenings. In Madhya Pradesh it is reported to live in cracks and crevices in the ground and it is also said to occupy the space between the roof and the ceiling of the less frequented forest rest houses (21). Normally it is a burrow-dweller often going head first into its bolt holes and remaining in that position till it needs to come out again. Once wedged in a rock crevice in this manner it is exceedingly difficult to extricate a monitor as it inflates its body thereby giving itself better purchase. The stories of monitor lizards being used to scale walls of forts could

Photo: T.N.A. Perumal

Common Indian Monitor (*Varanus bengalensis*) (See p. 58)

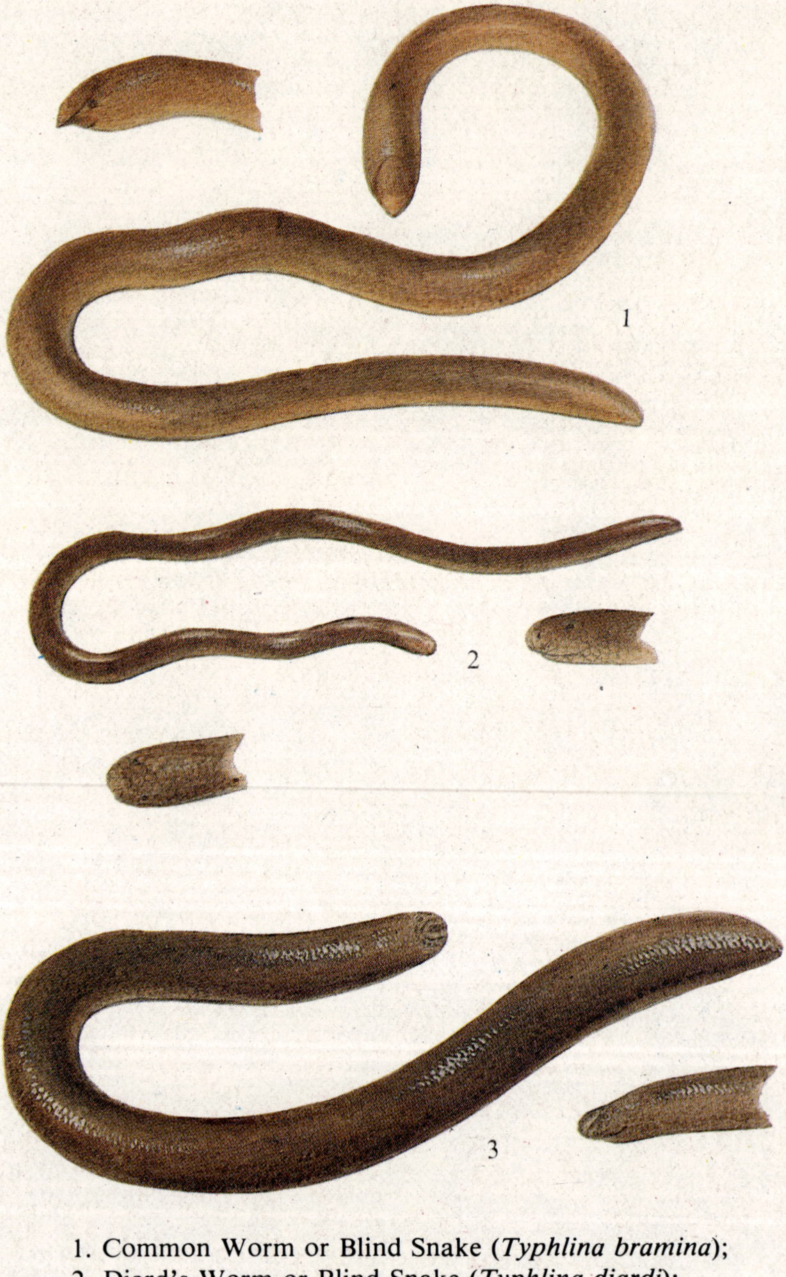

1. Common Worm or Blind Snake (*Typhlina bramina*);
2. Diard's Worm or Blind Snake (*Typhlina diardi*);
3. Beaked Worm or Blind Snake (*Typhlina acutus*)
(See pp. 64, 65)

be true, as a large monitor, once it is wedged in a hole, can very well support a person of light weight for a short time. When out foraging they move slowly along the forest floor with the tongue flicking in and out of the mouth like a snake's tongue. When necessary they can run at speed and they are also good climbers. Monitor lizards swim well and can remain submerged for a considerable time. When alarmed the monitor usually tries to escape notice by lying absolutely still. If seen and chased, runs with the tail raised at an angle of about 45° from the ground. A cornered monitor will demonstrate by rising on its forelegs, hissing, and lashing its tail. A captive, if handled carelessly, can inflict a painful bite with its recurved teeth which are difficult to dislodge.

FOOD. Carnivorous. Eats any animal it can overcome. Recorded food items include small mammals, birds, birds's eggs, eggs of crocodiles and small reptiles such as skinks, calotes, small turtles and snakes, fish, crabs and prawns, insects, arachnids, and carrion. It probably seeks its prey both by smell and sight. It has been seen removing eggs from bird-nests and young from squirrel-dreys. A monitor was seen climbing a tree, stalking and capturing a roosting bat. Fish and crabs caught in water are brought ashore and eaten. 21 scorpions are reported to have been removed from the stomach of a monitor.

BREEDING. The male during the breeding season apparently has a territory, as combats similar to the 'combat dance' of snakes have been recorded. The commonest manoeuvre in such combats is to stand up on the hindlegs, clasping the opponent firmly about the neck and shoulders and with a sharp sideways jerk of the head to knock it down, sometimes tossing it completely over. The struggle also includes biting on the nape of the neck without drawing blood. A combat which was watched went on for an hour and a half with occasional pauses (6, 9).

Eggs have been collected from mid April to October. The clutch size varies from 8 to 30 depending on the size of the animal, the larger females, as among other reptiles, lay more eggs. Size (average of 50 eggs), 49 × 38 mm (47 × 36 to 55 × 44 mm). Weight—average of 25 eggs—11.4 g (9.3 to 14.5 g). Eggs white, oval and soft-shelled. A female was seen digging a hole 28 cm in depth in which she laid 24 eggs. The hole was refilled and the soil tamped down with the snout. She then dug a few false pits near by before leaving the area (23). It is reported to lay in termite nests and to close the hole with leaves and rubbish. Incubation period 8 to 9 months. The newly hatched young are common at the beginning of the monsoon.

Three other species of monitor lizards occur in the subcontinent. The DESERT MONITOR, *Varanus griseus* (Daudin), inhabits arid sandy country in Pakistan and the North West of India. In colour pale sand grey with six or seven dark cross bands on body and proximal part of the tail. Distinguished from other monitors by its rounded tail. The YELLOW MONITOR *Varanus flavescens* (Gray) occurs in the Gangetic Plain from Punjab to Bengal and is said to be common in Bihar. In colour it is yellow or yellowish brown with broad red cross bands which are markedly evident during the rainy season but indistinct at other times of the year. The young are dark brown above with yellow spots or bars. Lips, throat, and belly with dark brown crossbars, otherwise yellow below. The yellow monitor has the nostril situated closer to the tip of the snout than to the eye. The WATER MONITOR *Varanus salvator* (Laurenti) is the largest among the Indian monitor lizards reaching a total length of *c.* 2.5 m. A large male 2.35 m in length weighed *c.* 14 kg. A female 1.80 m in length weighed *c.* 8 kg. The young are blackish above with yellow spots or ocelli arranged in transverse rows. The pattern fades with age. In the Indian region it occurs in Orissa, Bengal, eastern India, and Bangladesh. The distribution extends eastwards. It also occurs in Sri Lanka and the Andaman Islands. The most aquatic of Indian monitors, it is thoroughly at home in fresh as well as salt water. Frequents coastal estuaries searching for crustaceans and molluscs when the tide is out. It is as catholic in its tastes as other monitors. One specimen, caught in Burma immediately after it had been fishing for frogs in a rainwater pool, had eaten 40 frogs (71). It is fond of eggs and can be destructive to poultry. Readily climbs trees. Eggs 25 to 30 in number are laid in holes in banks or in tree-holes or in termite nests. Breeds at the beginning of the rains.

All monitor lizards are endangered by the trade in reptile skins. All are now listed in Schedule I, (completely protected species) of the Wildlife Preservation Act of 1972.

SNAKES

The Order Squamata of reptiles has two Suborders, the Serpentes or Snakes and the Sauria or Lizards. Snakes can be distinguished from lizards only by a combination of characters. Briefly, these are, the absence of fusion between the two halves of the lower jaw, which are united by ligaments only, and are therefore movable independently; the absence of eyelids, external ear openings, and of limbs. A few primitive forms have vestigial hindlimbs.

The elongate body of the snake is remarkably supple and is able to make twisting and other movements impossible for other vertebrate animals. The arrangement of the backbone does not give much latitude for vertical movements, and vertical undulatory movements are impossible for snakes. Lateral movements are the typical snake movement.

The scales on the body of the snake are imbricate, that is they overlap in the manner of tiles, and form patterns which are characteristic of the species. The skin, and the scales on it, are so constructed that stretching is possible to an enormous degree enabling snakes to swallow large prey. The skin is shed periodically and the shed skin is often unbroken, and maintains its form to a remarkable degree.

The mouth is armed with numerous teeth which are not embedded in sockets. There is a continuous succession of teeth, those that are lost being immediately replaced by successor teeth lying below. The teeth are recurved and serve to hold the prey, and thus assist in swallowing which is virtually done by the two halves of the jaw alternately walking over the prey and thus pushing it down the throat. Three types of fangs occur in harmful snakes. In the back fanged snakes, the last two or three teeth of the upper jaw are large and are grooved. The groove is connected by a duct to the poison glands. In snakes with fangs in the front of the mouth such as the Cobras, Kraits and the Vipers the groove has become a closed canal for the conveyance of the poison. The forked tongue in snakes is an organ of smell rather than of taste and serves to collect scent particles by its constant quiver and play. Snakes have no external ears and cannot hear noise carried through the air but they are able to feel through their jaws, vibrations carried through the substratum. The eye varies in size and effectiveness. In some snakes including the blind snakes, the eye is hidden beneath a head shield. When visible the eyes lack lids but have a transparent watch-glass-like shield beneath which they move. The pupil may be circular, vertical or horizontal. Depending on the species of snakes, the food varies from insects to large animals. Some feed on other snakes. The majority of snakes lay eggs with a white or yellow parchment-like shell. Parental care in the form of brooding is seen in many instances. Many species are ovoviviparous and give birth to live young. Snakes are widely distributed in the Indian subcontinent from the seas to near the snowline in the Himalayas.

SNAKE VENOM. Many species of animals have the capacity to produce venom.

The venom has a passive defensive function in frogs, toads and salamanders in which the venom glands are distributed in the skin. The venom apparatus reaches the highest developemet in snakes and is a weapon for capturing prey, defence and a digestive aid. The venom glands are actually specialised salivary organs and inoculation of the venom is through the canalized or grooved teeth. The salivary secretion of the harmless snakes is equally effective against their prey species. The venom not only immobilizes the prey but aids subsequent digestion of the animal tissues. Snake venom is a mixture, chiefly of proteins, varying in compostion from species to species. Studies so far undertaken indicate that the biological significance of snake venoms is primarily in their digestive role. Snakes cannot chew and mix the products of their salivary glands with the tissues of their prey. Instead they use a highly developed injection apparatus to apply digestive aids to their food. These powerful and concentrated enzymes are extremely poisonous. For instance the lethal dose of Russell's Viper venom for rabbit is 0.05 mg when injected intravenously. To enable the poisons to spread rapidly in the body, compounds in the poison break down the connective tissues and destroy blood vessels causing spread of erythrocystes and serum into the tissues. Necrosis at the site of a bite is due to the digestive properties of snake venom. A non-toxic component of the venom, not always present, liberates proteolytic enzymes into the victim's body and hastens putrefaction.

SNAKE BITE AND EFFECTS. It is not correct to assume that bites from harmful snakes are invariably fatal. Hospital case histories with definite identification of the biting snake show that in about half the cases there is little or no effect of the poison. This is understandable considering the fact that snake bite on man is a purely defensive reaction.

The local symptoms of poisoning are distinctive and consist of pain, immediate swelling and later blisters and necrosis. These symptoms vary according to the species. Severe pain is felt after a cobra or viper bite starting within a few minutes of the bite in the case of the vipers and reaching a maximum in about 12 hours. In the case of the cobra swelling starts about 1 to 3 hours after the bite and reaches a maximum in 24 to 48 hours. Swelling of the whole limb occurs in viper and cobra bites and the swelling is tender in both cases. Poisoning by Krait may not cause local swelling.

Discoloration of the skin around the bite occurs both in cobra and viper bites and blisters appear around the bite and extend up the body in the case of viper bite if a large amount of venom is injected. Local necrosis is invariable in cobra poisoning and may be extensive even if more serious effects of the poison do not develop.

The toxic effects of the poison in the case of cobra and krait poisoning are drowsiness commencing in 15 minutes to 5 hours in cobra and may be delayed up to 8 hours in krait poisoning; difficulty in opening the eyes and mouth, moving the lips and swallowing. Severe abdominal pain occurs in the case of krait poisoning. Limb weakness develops last. The victim is unable to sit up or hold the head up. The lower jaw falls and drools saliva, breathing becomes increasingly difficult.

The most significant symptom of viper poisoning is bleeding from the bite site, any sores on the body and internal haemorrhage. Heart sounds become faint.

Death occurs quickly in cobra and krait poisoning and is delayed in viper bites. If by chance the strike of the snake ruptures a vein, death may occur within 15 minutes in either case.

The treatment recommended is to reassure the person by a placebo injection if necessary. The site of the bite should be gently wiped and covered with a clean handkerchief. No incision should be made as it often introduces infection. Apply a firm but not tight ligature above the bite with a cloth or handkerchief. The patient should be taken to the nearest doctor or hospital for treatment. The polyvalent serum now available against the bites of cobra, krait, Russell's viper and Echis makes death from snake bite exceptional.

COMMON WORM OR BLIND SNAKE

Typhlina bramina (Daudin)

HARMLESS

LOCAL NAMES. Hindi *Do mukh ka samp, Telia sap*; Tamil *Sevi pamboo*; Marathi *Vala*; Gujarati *An-sap*; Oriya *Do-mundia sápá*.

SIZE. Grows to about 170 mm.

IDENTIFICATION. All *Typhlina* have a slender worm-like shape and undifferentiated body scales but distinguishing between species is difficult. *T. bramina* differs from all other *Typhlina* in the line of division (suture) of the nasal shield touching the preocular shield (shield before eye) instead of to the 2nd. labial (lip) shield. Body uniformly cylindrical. Head bluntly rounded. Eyes indistinct. Tail very short ending in a small sharp stiff point. Scales highly polished. There are 20 rows of costal (body) scales.

COLORATION. Brown or blackish brown above lighter below, snout, anal region, and end of tail pale. A pale buff (in life, flesh coloured) variety *arenicola* is known from the sandy tracts of Ramanathapuram Dt., Tamil Nadu.

HABITAT, DISTRIBUTION, and STATUS. Lives beneath the soil, or stones, or debris, and is seen only when these are turned over, or when flushed out of its subterranean burrow by rain. Throughout South Asia from Arabia in the west to S. China in the east and the Philippines in the south; Africa; Zanzibar; Madagascar; and the Indian Ocean islands; Mexico. The most widespread species of the genus, possibly carried unwittingly in man's effects.

HABITS. Usually solitary, but occasionally in large aggregations in rotting wood. An active burrower, its highly polished scales help its passage through the soil. A slow mover above ground it makes vigorous attempts to escape if dislodged from its hiding place or when handled, 'anchors' itself with its tail spine and wriggles restlessly in all directions. Often seen inside houses. It is preyed on by other snakes. Several have been recovered from the stomach of Kraits. Popularly supposed to enter the ear of people sleeping on the ground, hence the Tamil name: *Sevi pamboo* (ear snake).

FOOD. Larvae, pupae, and imagines of ants and other small insects. A snake in captivity ate caterpillar droppings.

BREEDING. Oviparous. Breeds from April to July in Assam. Eggs 2 to 7, white, and of the size of cooked rice. Believed to be parthenogenetic.

DIARD'S WORM OR BLIND SNAKE

Typhlina diardi (Schlegel)

HARMLESS

A much larger species than *bramina*; grows up to 430 mm in length and is common in West Bengal and further east. In colour, black or blackish above, paler below, or a uniform pale grey. In habits, it resembles the Common Blind Snake. Has 24 or 26 costal rows, rarely 22, or 28. It is reported to be ovo-viviparous and the clutch size varies from 4 to 14. Gravid females have been obtained from May to July. The young between 96 and 100 mm in length appear in September and are believed to attain sexual maturity at the end of 2 years when they are over 230 mm in length.

BEAKED WORM OR BLIND SNAKE

Typhlina acutus (Dum. and Bibr.)

HARMLESS

The largest among Indian worm snakes growing to a length of 600 mm and the only Indian species with a beaked snout. Brownish or blackish above, paler below, each scale with a paler centre. Habits similar to those of the Common Blind Snake, but the Beaked Blind Snake is believed to feed also on worms. DISTRIBUTION. Peninsular India, south of the Gangetic plain, but rarer south of latitude 16°N.

OCELLATE SHIELD-TAIL

Uropeltis ocellatus (Beddome)

HARMLESS

SIZE. Longest measured 530 mm.

IDENTIFICATION. Distinguished by the obliquely truncated tail with the terminal scute ending in two points, 17 costal (body) scale rows and ventral scale rows 185 to 234. A small compact snake with the head smaller in girth than the body and tapering to an obtuse point. Neck region similar in girth to the rest of the body. Eye small and contained in ocular or eye shield. Tail short and subtruncate. The females are usually larger with a longer body but the tail is relatively shorter.

COLORATION. Scales olivaceous brown or olivaceous green at edges, with paler centre. Whole body with numerous bright canary yellow spots grouped to form irregular chains across the back, incorporated in rather ill-defined dark crossbars, joining to form a yellow patch below.

HABITAT, DISTRIBUTION, AND STATUS. Inhabits the moist soft soils of dense forests of the Nilgiri, Anamalai and other hills in south India between 600 and 1000 m. Common in the Nilgiri Wynaad where 100 specimens were obtained in a year, all from the western slopes.

HABITS. An inoffensive, quiet, little snake living by choice beneath the soil or beneath stones and logs in the forest. Burrows easily in loose earth using its snout for the purpose.

FOOD. Feeds almost exclusively on earthworms.

BREEDING. Ovo-viviparous, 3 to 5 young being born at a time. The breeding season in the Nilgiri's apparently ends by July. Young at birth are 11 to 12 cm in length and grow another 5 to 7 cm in the first year of life.

The family Uropeltidae is restricted to the high rainfall forests of the Western Ghats of the peninsula of India and Sri Lanka. Several other genera of the family occur in the Western Ghats mainly south of 12°N latitude. The largest number of species occur in the south-west portion of the Peninsula in the Western Ghats and associated hills from sea level to 2500 m.

1

2

1. McClelland's Coral Snake (*Callophis macclellandi*);
2. Ocellate Shield-tail (*Uropeltis ocellatus*) (See pp. 110, 66)

1

2

3

4

5

6

1. John's Earth Boa (*Eryx johni*); 2. Russell's Earth Boa (*Eryx conicus*) (See pp. 69, 67)

Russell's Earth Boa

Eryx conicus (Schneider)

HARMLESS

LOCAL NAMES. Hindi *Thut thur samp, Do muha*; Gujarati *Bharampodi, Kodhio*; Marathi *Durkiya ghonas, Kakria*; Kannada *Itally havu*; Tamil *Man-nooly*; Malayalam *Mandally*; Telugu *Mondi poda*.

SIZE. Longest measured 2 ft 7 in. (*c.* 78 cm). Specimens over 2 ft. (*c.* 60 cm) are not common.

Females grow distinctly larger than the male. The male has a proportionately longer tail.

IDENTIFICATION. The dual association of small head scales, with ventrals so narrow that they are only twice the breadth of the last costal row, suffices to pronounce the snake an *Eryx. E. conicus* differs from the other Indian species *E. johni* in the *absence* of a 'groove' beneath the chin and angular transverse ridge on the rostral shield. A markedly stout snake; body short and heavy and tail short and tapering very rapidly so that it is conical in shape. The body narrows very gradually in both directions passing almost insensibly into the head with but slight indication of a neck. Head moderately elongate, rounded evenly from side to side; snout long, overhanging the chin; eye very small; pupil vertically elliptic; iris beautifully speckled with gold, nostrils slit-like and placed high on the snout; tongue pale at the base and blackish at the tip. At each side just above the vent is a small curved claw-like process directed backwards in the female which indicate the termination of the rudimentary hind limb. It is well developed in the male. The skin is rough on the back owing to the keeled scales; this roughness is very pronounced on the hinder part of the body and on the tail.

COLORATION. Underparts buff, uniform or with traces of mottling; flanks mottled brown, varying from light brown to deep chocolate; the mottling becomes coarser as it ascends the flanks; vertical bars of the ground colour pass up to the spine where they meet and form irregularly squarish blotches of the same colour running from the nape to the tail tip. Often bars on the two sides may alternate resulting in an irregular dark patchy confluent pattern. Head, light above, sometimes with dark speckling about the lips. A dark, irregular stripe from eye to the gape.

HABITAT, DISTRIBUTION and STATUS. Generally considered a snake of arid areas but it is by no means confined to such tracts. It is a very common snake in Malabar where the annual rainfall is about 381 cm and the soil supports a particularly luxuriant vegetation. Occurs from the base of the Himalayas to the extreme south of India and from Sind and Baluchistan in the west to Bihar and Bengal in the east; North Sri Lanka. A snake of the plains, but like many other species, ascends some distance into hilly regions and has been collected at Khandala (W. Ghats), Annamalais (Tamil Nadu) and in Naini Tal District (U.P.). A fairly common snake.

HABITS. Russell's Earth Boa is a dull, phlegmatic creature, but of uncertain temper. While some bite on the least provocation others when irritated sulk in a most determined manner, remaining quite motionless and refusing to offer any resistance. Like the rest of the genus it is an earth snake and burrows into loose soil easily and speedily. However, its life is by no means completely subterranean and a considerable period of it is spent either above the soil or in the most superficial layers, into which light is admitted, and by its stimulus the function of the eye is preserved. As a result this organ though small, is quite as well developed and vision seemingly quite as good, as in colubrines and other highly developed snakes. Specimens have also been frequently dug up from deeper layers of the soil. This snake is very frequently abroad by daylight and even in the midday glare of the hot weather often establishes itself beneath trees in south India where it lurks for the purpose of catching Palm Squirrels (*Funambulus* sp.) when they descend and wander about beneath the trees. With this object it partially buries itself and patiently awaits the chance of a squirrel coming within reach. The keeled scales of the body undoubtedly assist its concealment. As the snake noses its way through the surface soil, particles of earth lodge in the grooves formed between the keels on its back and serve to conceal its serpentine form. The normal movements of the snake are laboured, and slow. However when annoyed as when touched it leaps with rigid body about a centimetre up and off the ground and slithers along the ground in short, sharp, jerky side to side leaps, maintaining the rigidity of the body. Like other Indian representatives of its family, it sheds its skin about 4 times a year at rather irregular intervals.

FOOD. Feeds largely on small mammals particularly rodents. Frogs are occasionally taken but reptiles are not usually acceptable, though instances are recorded of other snakes being killed by them. Like other boas it kills its victims by constriction and the strength of its body is such that a squirrel or mouse is killed in a few seconds, and until life is extinct it does not commence swallowing. Ground feeding birds such as mynas, hoopoes and babblers often fall victim. In the Dera Ismail Khan area (Pakistan) it was noted to subsist largely on quails during the migratory season.

BREEDING. In southern India mating occurs in November. It is ovo-viviparous. A female killed in December contained 6 largish eggs, about 2.5 cm long. Another is recorded with 16 eggs in January. A female was collected with 3 young in August and another with 5 young in June. A female in captivity delivered seven young of length *c.* 8 cm in June.

JOHN'S EARTH BOA
Eryx johni (Russell)

HARMLESS

LOCAL NAMES. Hindi *Do muha*; Marathi *Do tondya*; Gujarati *Chákalan, Andhali chákalan*; Tamil *Iruthalai pamboo*; *Mannooli*; Kannada *Italqi havoo*; Malayalam *Mandalli*.

SIZE. Adults rarely exceed (90 cm): longest measured 4 ft 2 in. (*c.* 1.25 m), wt 2½ lbs (*c.* 1 kg).

IDENTIFICATION. Distinguished from Russell's Earth Boa (*Eryx conicus*) by the presence of a mental (chin) groove, the pronounced angular ridge on the muzzle, the blunt tail and the larger number of costal scales, over 53. A stout, heavy and muscular snake more or less uniform in girth from head to tail with little constriction at the neck. A groove along the spine. Body scales small, smooth. Head scales a little larger than back scales. Muzzle broad with a very pronounced horizontal ridge. Upper jaw overhangs the chin. Eye small; pupil vertically elliptical; iris spotted with ruddy gold. Nostril high on the snout, slit like and placed between two enlarged shields. Tongue yellowish basally, black at the tips. A longitudinal furrow (the mental groove) beneath chin. Tail short, stumpy, rounded at its end and in general form very similar to the head. Claw like rudimentary limbs as in *E. conicus*.

COLORATION. Adults uniform light brown, ruddy-brown, or dark olivaceous-brown with or without fine dark reticulations arranged to form crossbars. Markings when present conspicuous posteriorly. Belly lighter than back and often mottled with black. Young brick-red or sandy-red, individual scales being pale buff with very fine ruddy reticulations. Back with dark blotches or ruddy-brown crossbars. Belly buff heavily dappled with darker tones.

HABITAT, DISTRIBUTION, and STATUS. Widely distributed in the plains of the Indian Subcontinent. In hills up to about 600 m elevation. Recorded as common in some areas.

HABITS. In general habits resembles *E. conicus* but appears to be of a gentler temperament being a particularly inoffensive creature. It is a deft and speedy burrower, nosing its way into the earth by using the transverse ridge on its snout as a digging implement. Its movements above ground in keeping with its heavy body and reduced ventrals are slow and clumsy.

FOOD. Mainly mamalian. Rats, mice and other small rodents are killed by constriction.

BREEDING. Very little is known of its breeding habits. Ovo-viviparous; six to eight young are born at a time.

MISCELLANEOUS. In South India, an age old belief is that the bite or a lick from this snake can cause leprosy! A belief in the Punjab is that if it bites any one, the same person will be bitten on each succeeding anniversary by the same snake which, however, is only visible to its victim. That the snake has two heads is a belief held all over the country as the common names suggest. It is also thought that the heads are alternatively in action every six months. Snake charmers who carry this species often mutilate the tail, making markings to suggest eyes and cutting a transverse incision at the tip which leaves a scar suggesting a mouth.

Indian Python (*Python molurus*) (See p. 71)

Trinket Snake (*Elaphe helena*) (See p. 74)

INDIAN PYTHON

Python molurus (Linnaeus)

HARMLESS

LOCAL NAMES. Hindi, Gujarati, Marathi *Azgar*; Tamil *Periya pamboo, Malai pamboo*; *Kaloodai viriyan*; Malayalam *Malam pamboo*; Telugu *Pedda poda, Condasella*; Kannada *Heba havoo*; Sinhalese *Prinbera*; Burmese *Sa-ba-ohn*; Bengali *Mayal, Ajagar*.

SIZE. Longest measured 19 ft 2 in. (*c.* 5.85 m) (Cooch-Behar, West Bengal); other large snakes on record: 19 ft (5.7 m) Sri Lanka, Travancore (Kerala), Assam, 18 ft (5.48 m) (Kerala, Assam, W. Bengal). One specimen weighed 200 lbs (90.7 kg).

IDENTIFICATION. The Python is easy to identify. The sensory pits on the rostral (snout shield) and first two labials (lip shields) distinguish it from all other Indian Snakes. Costals 58-73, ventrals 245-270. Caudal 60-72 pairs. A massively built snake, its girth relative to its length considerably exceeds that of all other snakes, except its own close allies. Rounded in outline and thickset in the middle, tapering towards head and anus. Scales smooth, and when the snake is in good condition, glossy. Neck distinct. Head flattened with a long snout. Nostrils large, directed upwards and situated high on the snout. Rostral and first two labials with sensory pits. Eyes small, pupil vertical, iris flecked with gold. Chin with mental groove. Tail short and prehensile tapering rapidly. Rudimentary hind limbs, as curved claw-like processes on either side of anus, more highly developed in male.

COLORATION. Ground colour greyish, whitish or yellowish in adults and in the young often a very pretty shade of pink. A dark streak from eye to nostril in young may or may not be present in adult. A conspicuous dark, oblique band, from eye to gape. Lower lip often mottled. On the back of the head and the nape is a large lance shaped mark with a pale centre often fading anteriorly in the adult. The body with a series of large, roughly quadrate patches from neck to tail dorsally. The patches are centrally of the same colour as the general body colour and are broadly outlined with black or blackish. Two, smaller rows of marking occur to the side of the median row. Ventrally dirty whitish or yellow. In a freshly moulted specimen the dorsal patches are highly iridescent giving bluish and amethystine hues in certain lights.

HABITAT, DISTRIBUTION, and STATUS. Normally a jungle dweller occurring in dense as well as in open forests with rocky outcrops. In the absence of forests occurs in rivers and jheels. Two races occur, *Python m. molurus* with 6th or 7th labial touching eye and lance-shaped mark on top of head distinct only posteriorly; Sri Lanka and peninsular India up to Sind in the west and Bengal in the east and *Python molurus bivittatus* with labials separated from eye by sub-oculars and lance shaped mark distinct on head; Indo-Chinese subregion. Not uncommon.

HABITS. A lethargic and slow moving snake even in its native haunts, exhibiting little, if any, timidity, rarely rousing itself seriously to escape even when attacked. Pythons also have a peculiar method of movement, a rectilinear progression, the body moving in a straight line like a millipede, waves of motion of the ribs can be seen in quick succession. Pythons climb well in search of prey and often hide among the branches of a tree. They are quite at home in water; swim deftly and strongly when necessary, but usually remain near the bank partially or wholly submerged except for the tip of the snout. They can, when needed, remain completely submerged for many minutes. In the backwaters area of Kerala, Pythons are reported to arrive in numbers during the rains to live on the rats infesting the coconut plantations and the fish occurring abundantly in the lagoons (43). Diurnal and/or nocturnal depending on the amount of disturbance from man in their environment. In north India and perhaps in the hills in the south, the Python hibernates during the cold season in any convenient retreat such as a tree hole or hole in a bank or under rocks or grass heaps. Often several hibernate together (91). A specimen in captivity was sluggish and somnolent for 113 days and its temperature dropped from 82°F to 73°F.

FOOD. Feeds on mammals, birds and reptiles indiscriminately but seems to prefer mammals. Stomach contents reported: frogs, toads, monitor lizards, wild ducks, peafowl, poultry, rats, hare, porcupine, langur, jackal, mousedeer, hogdeer, chital, sambar fawn, barking deer, chinkara, or Indian gazelle and leopard. Instances of fruits in the stomach are apparently the food of ingested prey. Pythons which lie up in jheels and other waterlogged areas have little difficulty in capturing prey, largely water birds and animals which live near water or come to drink. They also lie in wait near jungle trails. Heronries, nesting colonies and large bird roosts usually have attendant python or pythons which take their toll at night. In one instance where a python's capture of a chinkara was observed, the animals rolled for about 45 m after the initial strike before the snake could subdue the antelope (38). In captivity accepts any mammal, bird, or reptile that is offered and not necessarily live prey. Live prey are constricted and killed. Aroused to activity on sighting the prey, the snake advances often with quivering tail and lunges with open mouth. One or two coils of the body are thrown around the prey holding it in a vicelike grip. The prey unable to breathe, soon succumbs and is swallowed head first. Prior to swallowing it is 'smelt' all over with the tongue. The head is seized and the movable lower jaws alternately inch forward and assisted by a copious flow of saliva gradually swallow the prey. During the slow process of swallowing the glottis or wind pipe extends an inch or two to assist breathing. The snake is disinclined to move after a heavy meal and if forced to move, hard parts of the prey such as horns may tear through the body wall (81). The period between meals varies with the condition of the snake, time of the year, and the size of the prey. Records indicate that in captivity small creatures like rats and crows are digested in about 8 days and a goat in about 3 weeks in summer. Healthy snakes digest all but the hard structures like feathers, teeth, beak, claws, horns and scales, which at excretion by the snake often retain to a wonderful degree their relationship in life. In captivity the snake often refuses food for considerable periods without any ill effect. Such a fast lasting over two years has been recorded (52).

BREEDING. Mating occurs during the cold season in December, January and February when pythons in north India are in hibernation, and eggs varying in number from 8 to 100 (largest clutch 107) are laid 3 to 4 months later in the hot weather months of March, April, May and June. Eggs measure 12 × 6 cm and are soft, white in colour and equally domed at both poles. The female broods the eggs by coiling around them. The body temperature is higher than normal and the brooding snake keeps it constant by shivering when the air temperature makes it necessary. The mother takes no further interest in her brood after hatching which occurs 58 days after laying. Females living alone in captivity often lay sterile eggs (49).

Hatchlings in one brood measured 73 cm. Growth is fairly rapid in the first few years and they breed at about 11 ft (3.35 m) length when they are approximately 5 years of age. Pythons are long lived, and have lived in captivity for over 22 years. Growth is a continuing process though at a slow rate. In captivity however, some may grow little, if at all, after attaining a length of 3.6 m. Sloughing in healthy snakes occurs five to six times a year, except during hibernation.

GENERAL. In some parts of India and further east the python is eaten. The flesh is stated to be 'tough and too sweet'. The skin, as of other snakes is commercially used. An albino python has been recorded in W. Bengal (48).

RETICULATED PYTHON *Python reticulatus* (Schneider)

HARMLESS

The Reticulated Python (*Python reticulatus*) is distinguished from the Indian Python by the presence of pits on the 3rd and 4th lip shields and larger number, 297 of costal (body) and ventral shields. Occurs in our region only in east Assam, Nicobar Islands and Bangladesh. The longest snake in existence, the authentic recorded length is 28 ft (8.4 m) (34) with weight 250 lbs (113. 6 kg). The specimen a female, kept in captivity in the Hagenbeck collection in Germany, laid 100 eggs in October of which 45 hatched out 2½ months later in January. Young at birth 61 to 76 cm in length. A Reticulated Python hatchling born in captivity with a length of 86 cm grew to 4 m in three years. In the first year it grew an additional 1.57 m reaching a length of 2.40 m (90). Reported to be once abundant in Bangkok living on the domestic animals of the city; said to prefer small mammals but recorded to take large animals up to the size of a full grown wild boar.

TRINKET SNAKE

Elaphe helena (Daudin)

HARMLESS

LOCAL NAMES. Marathi *Taskar*; Tel. *Mega-rekula-poda*; Singhalese *Mudu karawala*

SIZE. Specimens over 4 ft unusual. Longest measured 5 ft 3 in. ♀ 1640 mm, ♂ 1100 mm (Smith).

IDENTIFICATION. Scale rows at mid body 25-29. Anal 1. Two or three labials touching eye.

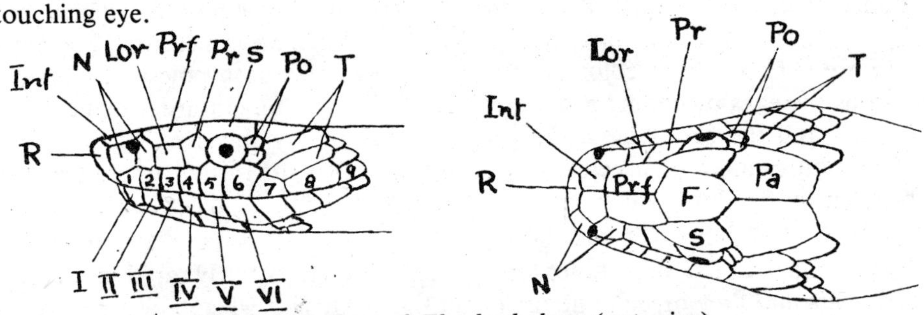

1 Head Scalation of *Elaphe helena* (nat. size)

Key character: 5th, 6th labials (upper lip shields) touching eye.

COLORATION. Above, brown of varying degrees of intensity. Yellowish or whitish on lips. An oblique streak from eye to lip. Some or all sutures on both lips dark streaked. Forebody for a variable length beautifully ornamented with ocellated crossbars of a pattern peculiar to this snake. In specimens from Western Ghats anterior crossbars are connected by festoons on the sides. The patterns become modified in time and gradually disappear. Usually on nape a conspicous V-shaped black mark, or two lines either parallel or occasionally meeting to form two or three crossbars on back. A broad dark stripe extending to tip of tail gradually replaces crossbars on sides of body.

HABITAT, DISTRIBUTION, and STATUS. Usually found in or at the periphery of forests. Occasionally encountered near human habitations. Peninsular India and from Sind in the west to Assam in the east. Himalayas, Sri Lanka. Common between 500 m and 2000 m. Rarer in the plains.

HABITS. Truculent when annoyed. The neck is compressed, spine arched and throat distinctly pouched. Erects and throws the forebody into broad sigmoid curves which are straightened in the act of striking. The scales of the body when thus excited, separate and reveal the pinkish skin of the intervening skin. Strikes repeatedly and upward. Prefers a mammalian diet but takes lizards and even snakes when hungry. It has the habit of coiling around its prey but apparently does not constrict. The young hatch prior to June. The smallest specimen obtained measured 287 mm and it is thought that they nearly double their length in the first year.

1

2

3

4

5

6

7

1, 2. Fasciolated Ratsnake or Banded Racer (*Argyrogena fasciolatus*); 3-4. Gray's Ratsnake or Glossy-bellied Racer (*Argyrogena ventrimaculata*); 5-7. Copper Head (*Elaphe radiata*) (See pp. 78, 79, 75)

1

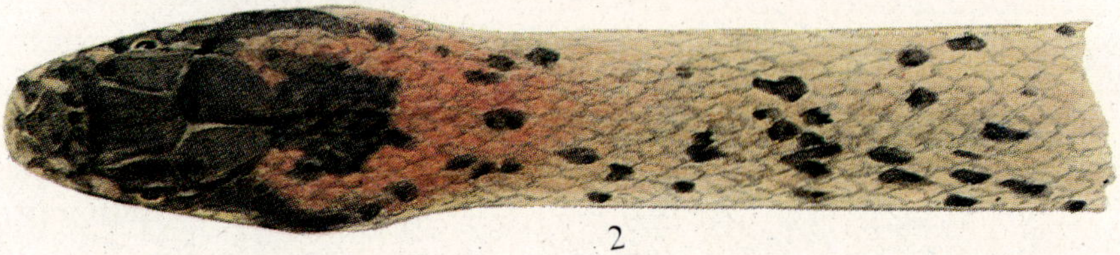

2

3

4

5

6

Royal or Diadem Snake

1-3. *Spalerosophis diadema* var. *atriceps*; 4-6. *Spalerosophis diadema* var. *diadema* (See p. 80)

Copper Head
Elaphe radiata (Schlegel)

HARMLESS

LOCAL NAMES. None recorded.

SIZE. From 1.5 to 1.80 m. Longest measured 2.10 m.

IDENTIFICATION. The following combination, 19 body scales (costals) at mid body, median rows keeled, anal 1, ventrals 224-250, caudals 83-103 and a black transverse mark on the back of the head distinguish the species. A moderately large, handsome snake with distinctive marking. Head elongate. Iris golden or golden brown. Nostril occupies whole suture between nasals. Tongue pale at base, black tipped. Body comrressed, keeled.

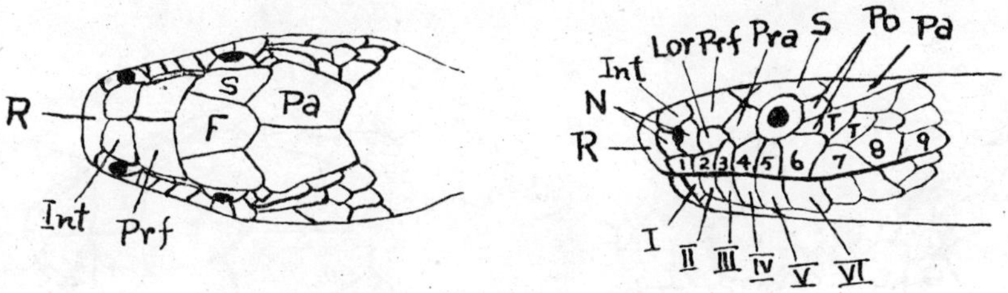

Head Scalation *Elaphe radiata* (nat. size)

For key to abbreviations see Glossary

COLORATION. Head copper or dull orange. A transverse black stripe across head with black streaks leading to the eyes and two backwards. Two stripes below eyes. Body yellowish, brownish, ruddy brown, or leaden grey anteriorly adorned with longitudinal black stripes, usually three on each side, median may or may not be connected to the black collar. Posteriorly the stripes disappear; belly white or pale yellow mottled with grey. Skin chequered black, blue-grey and bright yellow anteriorly, visible when the snake inflates itself under excitement.

HABITAT, DISTRIBUTION, and STATUS. Prefers open country near jungles and fields, gardens near villages. Northeast Orissa, Bengal, Eastern Himalayas, Assam and eastwards. Not uncommon.

HABITS. Diurnal, an active, intrepid snake, when annoyed erects forebody, compresses the neck forming a pouch in the throat and strikes. Apparently feeds exclusively on mammals, particularly rodents.

Gravid females have been obtained from April to July; eggs number 5 to 12, once 23 recorded. Largest measured 5 cm.

DHAMAN or COMMON RATSNAKE

Ptyas mucosus (Linn.)

HARMLESS

LOCAL NAMES. Hindi, Bengali, Marathi *Dhaman*; Kannada *Kere*; Tamil *Sarai pamboo*; Malayalam *Chera*; Singhalese *Garandiya*; Telugu *Jeri potoo*; Assamese *Machoa gom, Gola samp.*

SIZE. Majority of adults vary from 1.65 to 2 m but specimens up to 3.52 m (11 ft 9 in.) have been recorded. Males usually longer than females.

IDENTIFICATION. Costals 17 to 19 two head lengths from head; 17 to 16 mid body and 14 or 12 two head-lengths before vent. Head rather elongate, eyes large and lustrous. Nostrils large occupying the whole depth of the suture between nasals. Neck distinctly constricted. Body robust, compressed, tapering towards both ends. Tail cylinderical about one-fourth total length.

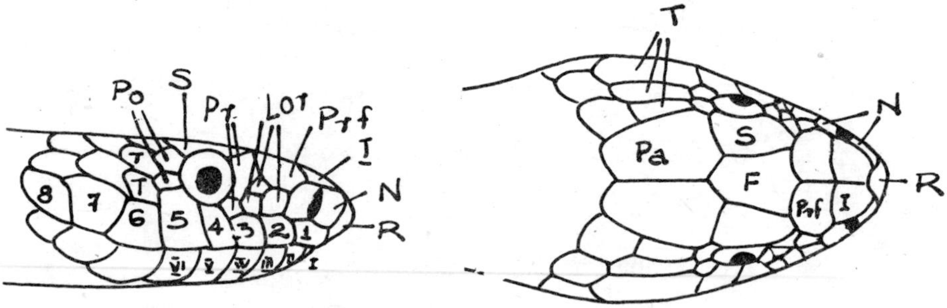

Head Scalation of *Ptyas mucosus* (nat. size)

For key to abbreviations see Glossary

COLORATION. Dorsally olivaceous brown, sometimes as dark as sepia or a light mustard yellow. Uniform or scales on the posterior part irregularly margined with black forming a reticulate pattern with a tendency to form crossbars. Lips and ventral scales margined with black. Belly greyish white, dirty white or yellowish. Skin blackish dorsally mottled with fawn or whitish transverse streaks, hidden by the scales. In young, however, light bluish grey irregular crossbars are usually conspicuous anteriorly.

HABITAT, DISTRIBUTION, and STATUS. Essentially a snake of the plains but has been recorded up to 1800 m. Throughout the Indian subcontinent, Sri Lanka and Burma. In the west extends to Afghanistan and Turkestan. In the east to south China. Also reported from Java and Sumatra. A common snake in all parts of the country.

HABITS. Diurnal but in populated areas may not be commonly seen out during the day. Takes readily to water and swims vigorously with the head well above the water. Dives with equal facility. Equally adept at climbing and is often seen on trees, has been reported to jump down from a height of 6 m when disturbed

(17). Normally tries to escape when sighted but if provoked and brought to bay attacks with courage and determination. When infuriated retracts and slightly erects the head and body into an S, compresses the body and with spine arched and the throat markedly pouched, strikes upwards. Often when thus demonstrating produces a peculiar mewling sound like a cat at bay (78). The tail tip is also rapidly vibrated (31,3).

FOOD. Eclectic in diet, devouring almost anything that chance brings within its reach but shows a marked preference for frogs and toads. Quarry once captured is swallowed at once and inoffensive creatures like frogs may reach the stomach alive enough for their suppressed cries to be distinctly audible and remarkable as it may seem, it is a fairly common event for frogs, when rescued immediately, to recover sufficiently to hop away, after a few minutes. When necessary the snake can exert considerable pressure to hold down its prey and can crush to death by pressing the prey against the ground with its body. Rats though preyed upon are not nearly as staple an article of diet as suggested by the snake's popular name. Lizards, birds and other small vertebrates are taken when available and the recorded food includes, geckos, toads, frogs, young pond turtles, nestling birds, skinks, agamid lizards, bats and snakes.

BREEDING. The breeding season varies with the climate. In the Himalayas and the higher hills of the south young usually appear from May to August, rarely in September. In the plains it mates in May-June, lays eggs in August-September and young are born from late September to January. Eggs in a clutch vary from 6 to 14 in number. Glazed white in appearance, equally domed at both sides and with a crisp, thick, parchment-like shell. Sticky when voided and adhere to each other. The female coils up with the eggs. Eggs measure 41-60 × 25-32 mm. Hatchlings measure 371 to 472 mm. One grew 3 cm in 20 days, though it had taken no food. The young apparently double their length in a year and continue this rate of growth the second year when they are about 125 mm in length. Sexually mature when they are about three years old.

The Dhaman has lived in captivity for over 11 years.

An unusual behaviour, which has not been recorded in any other species of Indian snakes is the 'Combat Dance', between males which twine around each other on the ground as well as when half erect. Most of the records of mating in literature relate to this behaviour (53, 4, 58, 66).

MISCELLANEOUS. The Rat Snake used to be eaten in some parts of Malabar in Kerala, the snake being de-scaled by passing it tail first through a hole in a coconut shell. The head is discarded and the cooked flesh is served as "Parambu wala or Kara wala" = Land Cat fish (72). There is a persistent belief in many parts of the country that the rat snake is the male of the Cobra. Another erroneous belief is that it is addicted to sucking the teats of cows.

FASCIOLATED RATSNAKE or BANDED RACER
Argyrogena fasciolatus (Shaw)

HARMLESS

LOCAL NAMES. Telugu *Mooni paragoodoo* (Vizag Area. Russell); Marathi *Nagin.*

SIZE. Largest measured 1.26 m (Smith).

IDENTIFICATION. Costals 21:23:17; anal 2, supralabials 8, 3rd divided and 4th and 5th touching eye. Rarely 4th divided and 4th, 5th and 6th touching eye. 5th or 6th highest touching temporal. Body elongate, fairly robust, cylindrical, tapering slightly at the neck, more markedly behind. Head moderately depressed, broadest between eye and neck. Upper jaw projects rather prominently. Tail one-fourth or one-fifth body length.

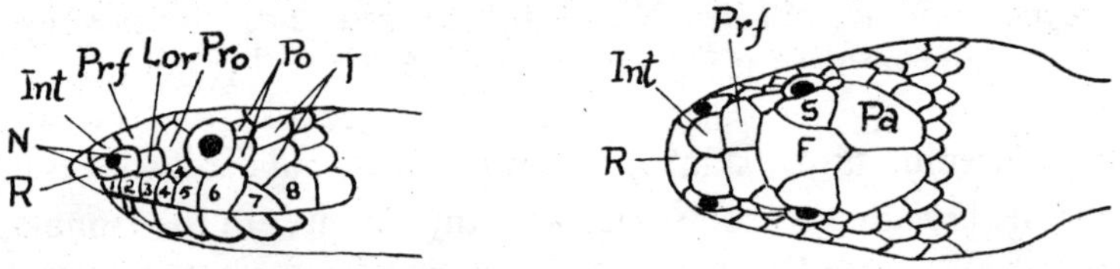

Head Scalation *Argyrogena fasciolatus*

Key characters: Supralabials (upper lip shield) 3rd divided, 4th, 5th touching eye, rarely 4th divided and 4th, 5th, 6th touching eye.

COLORATION. Brown or olive-brown varying in intensity from yellowish to a deep rich brown. Young beautifully ornamented with black and white crossbars conspicuous anteriorly and gradually disappearing before, at, or some distance behind mid body. Head usually unmarked. Belly white or yellow, greenish yellow in the very young.

HABITAT, DISTRIBUTION, and STATUS. Frequents jungle tracts, occasionally in urban areas and open places. Peninsular India, and up to Sind in Pakistan north to Himalayas and in the east to West Bengal. North Sri Lanka. Comparatively common in the Konkan and southwest Peninsula, rare elsewhere.

HABITS. A plucky and aggressive snake when molested but individual temperament varies. Becomes tame in captivity and can be handled without difficulty. When alarmed erects the forebody and flattens the body behind the neck like a cobra and it is often mistaken for one. In the Konkan region of Maharashtra it is believed to be the female of the cobra.

Feeds on small mammals (rats have been reported in the stomach) and amphibians.

BREEDING. Breeding habits little known. A gravid female had 5 eggs in early stages of developement in January. Young have been collected in July.

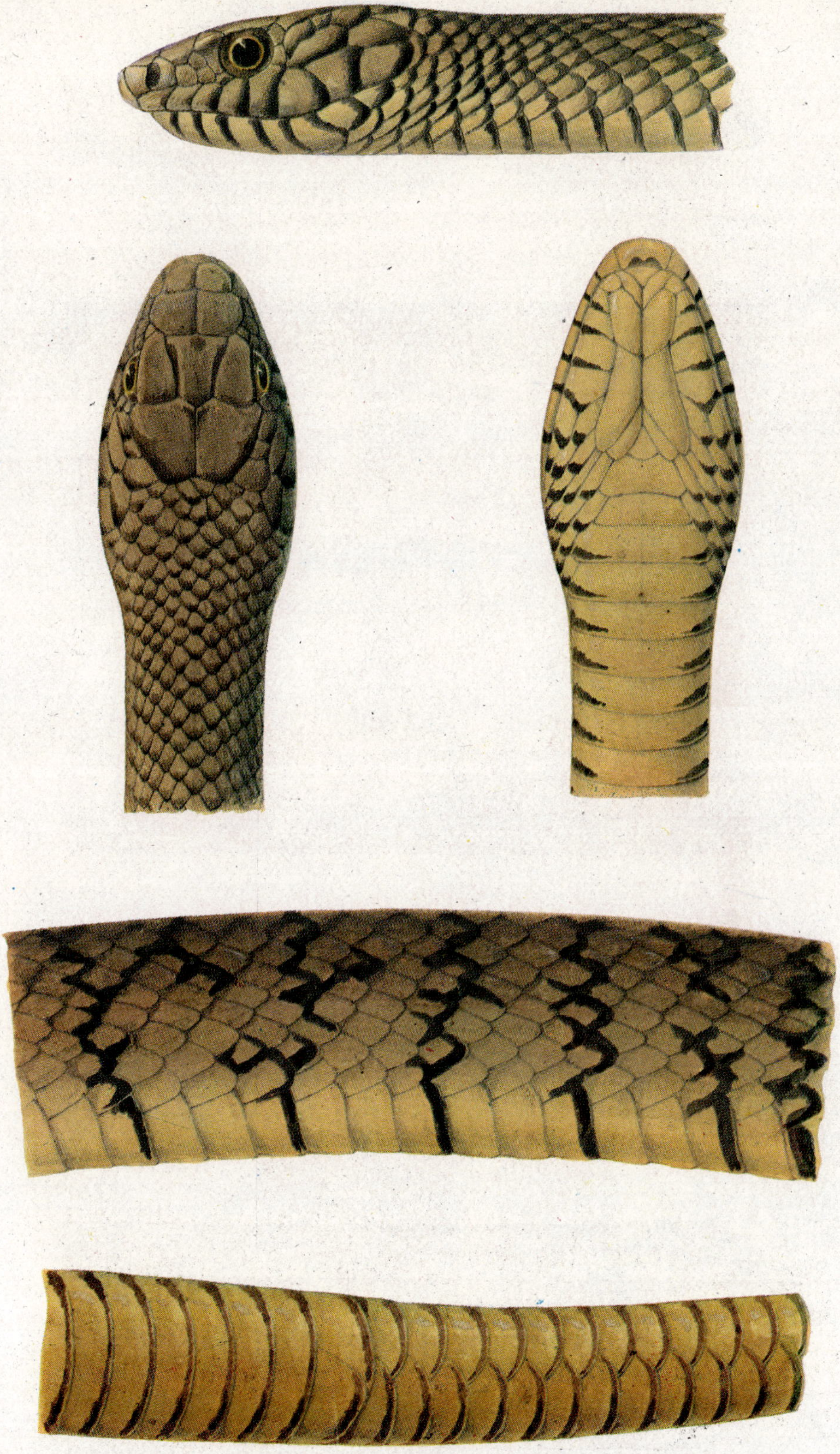

Dhaman or Common Ratsnake (*Ptyas mucosus*) (See p. 76)

1

2

3

4

5

6

1-3. Common Kukri Snake (*Oligodon arnensis*); 4-6. Ladder-back or Light-barred Kukri Snake (*Oligodon albocinctus*)
(See pp. 82, 84)

Gray's Ratsnake or Glossy-bellied Racer

Argyrogena ventromaculata (Gray & Hardwicke)

HARMLESS

LOCAL NAMES. Arabic *Dawaid-al-khail*; Urdu *Sagi*

SIZE. Adults usually 90 cm to 1.20 m in length. (1.28 m, Smith)

IDENTIFICATION. Costals in 19: 19: 15 or 13 rows. Anal 2. Supralabials 9, 4th divided, 4th, 5th and 6th touching eye. Rarely 8, 3rd divided and 3rd, 4th and 5th touching eye. A graceful snake with smooth, round, elongate, gradually tapering body with the tail more than one-fourth total length. Head moderately narrow, nostril occupies two-thirds suture between nasals.

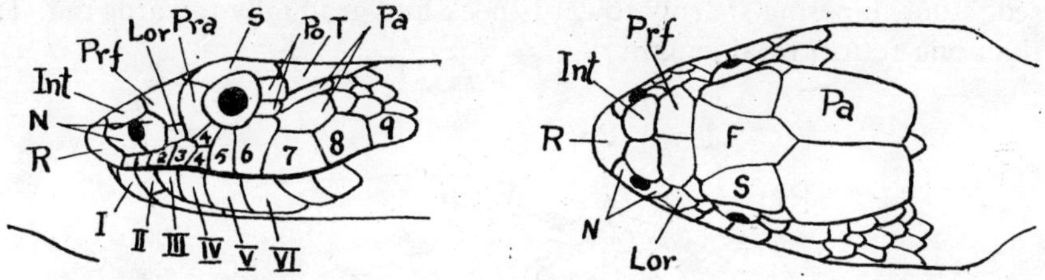

Head Scalation *Argyrogena ventromaculata*

Key Characters: Supralabials (upper lip scales) 9, 4th divided, 4, 5, 6 touching eye. Rarely 8, 3rd divided and 3, 4, 5 touching eye.

For key to abbreviations see Glossary

COLORATION. Dirty yellow, olive-green, olive-brown or grey above. Anteriorly marked with crossbars or spots or both. Bars may alternate with spots or bars may be replaced with five or three rows of spots, with a narrow crossbar. Head with a blackish spot on the lores, a black oblique streak below eye, another from the temporal area to gape. A band between eyes. A crossbar, and one or two stripes on nape. Belly saffron-yellow or paler.

HABITAT, DISTRIBUTION, and STATUS. Inhabits mainly stony hillsides, open or cultivated land. Occasionally found in congested urban areas. Northern India-Bombay-Poona to Rajasthan, W. Himalayas, westwards to the Middle East. Not uncommon.

HABITS. An active snake. Usually seen basking in the open close to cover into which it withdraws when disturbed. Feeds largely on lizards. Hibernates in winter.

BREEDING. Females obtained in early summer were gravid and young probably hatch in September when 30 to 33 cm long specimens have been collected. About 9 eggs are laid. Has lived in captivity for over five years.

ROYAL or DIADEM SNAKE

Spalerosophis diadema (Schlegel)

HARMLESS

LOCAL NAMES. Hindi *Rajitbansar* or *Rajitbans* (Rajasthan); Gujarati *Kevadiyo sap*.

SIZE. Longest measured 2 m.

IDENTIFICATION. Prefrontals in a double row of small scales, instead of the pair usual in Indian snakes. Costals of midbody 25 to 33 (27 to 31 usually) (29 or 31, rarely 27 or 33, Smith). Head longish oval. Neck distinct. Eyes rather small, pupil round, iris gold, often tinged brownish or reddish. Body compressed, stout, tapering sharply towards neck and gradually towards tail. Tail less than one-fourth total length.

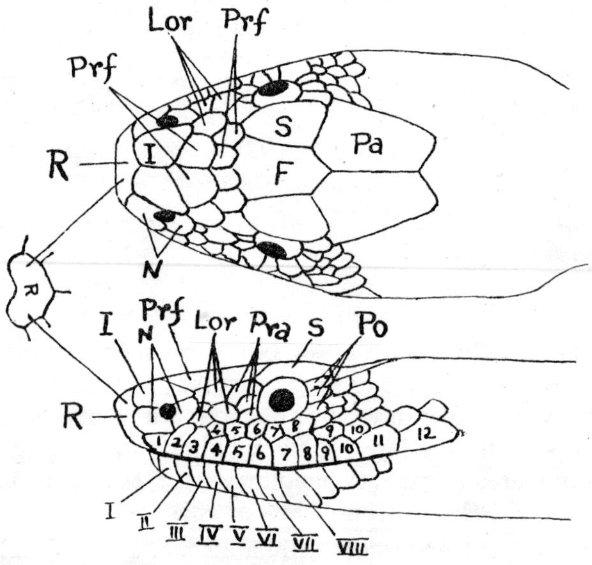

Head Scalation *Spalerosophis diadema* (*c.* 2)

Key Character: Prf. - Prefrontal in a double row of small scales, instead of the pair usual to Indian snakes.

For key to abbreviations, see Glossary

COLORATION. Two varieties: *diadema* light brown or fawn, with three sets of large dorsal spots, median row roundish or rhomboidal or as short transverse bars, from nape to tail. Alternate with smaller spots of the lateral series. Head light brown, spotted or mottled with darker markings. Often a band between eyes and a quoit-like mark on parietals (hence *diadema* = Crown), connected to each other or separate or may have two or three short stripes running backwards. Belly white. Markings may become obscure with age. var. *atriceps*: Buff, pinkish buff or pale brown, lighter on the flanks, which may be citron-yellow. Some scales deep claret, irregularly disposed. Head and neck strawberry scarlet or scarlet on neck, claret on head, merging or sharply divided. Belly uniform rose pink, mottled laterally. An apparently melanistic form, black with markings obsolescent. The two races are also considered as two distinct species.

HABITAT, DISTRIBUTION, and STATUS. Inhabits semi-arid and arid areas. Uttar Pradesh westwards to Algeria in North Africa. Eastern limits in India: Ganges (Allahabad, Fatehgarh). Southern limits roughly a line drawn from Allahabad to the south of Cutch. In Chitral considered to be the commonest snake between 1500 and 2000 m.

HABITS. In Chitral, Wall found them in stony country. Hibernates in crevices among stones in winter and as the year advances and the sun gets hotter emerges for basking and precipitately retires into its stony fastness at the least sign of danger. Feeds largely on rodents, climbing trees when necessary for this purpose.

BREEDING. Breeding habits little known. It is thought to lay eggs during May-July. The smallest specimens were obtained in October, November, February and March. Length 40.64 to 51.43 cm.

COMMON KUKRI SNAKE

Oligodon arnensis (Shaw)

HARMLESS.

LOCAL NAMES. Marathi *Gargar*; Tamil *Pul viriyan*; Telugu (Vizag Dt.) *Sanka*. The species name *arnensis* is derived from Arni, the town in South Arcot, Tamil Nadu where Russell obtained the specimen he illustrated. The name 'Kukri Snakes' for members of the genus *Oligodon* is suggested by the broad blade-like character and peculiar shape of the posterior maxillary teeth of snakes of the genus.

SIZE. Longest measured 66 cm. Specimens exceeding 60 cm very unusual.

IDENTIFICATION. Distinguished from other snakes of the Indian plains by the scale rows 17:17:15 and 4 or 5 infralabials. From other Kukri Snakes by the divided anal shield, presence of loreal, 7 supralabials and more than 40 caudals. Body cylindrical, short, smooth and even sized. Neck slightly indicated. Head depressed, snout short and blunt. Tail short, somewhat compressed basally and one-sixth to one-seventh of the total length.

Head Scalation *Oligodon arnensis* (× 1½)

Key Characters: Distinguished from snakes other than Kukri snakes by 4 or 5 infralabials (lower lip scales). From other Kukri snakes by the loreal (Lor) shield and 7 supralabials (upper lip scales)

For key to abbreviations, see Glossary

COLORATION. Ground colour brown of varying intensity, lighter on the flanks. Occasionally with a ruddy or purplish tint. Back with black bars, narrowly but distinctly outlined with whitish or pale yellow. Breaks up on the flanks into streaks. The number of bars on the body vary. According to M.A. Smith, Sri Lanka specimens have 13-18 on body 3-6 on tail, Indian specimens south of 20°N. latitude 18-30 on body, 4-16 on tail, Head with three conspicuous black marks, often narrowly bordered with white or yellow. Anterior mark crescentic, passing in front of eye to below them. Median and posterior, arrow-shaped, the apex of the former reaching the frontal and the arms, behind the gape. While the apex of the latter reaches the parietals and the arms, the sides of the neck. Below, pearly white, often spotted with black. In very old specimens the head markings tend to disintegrate and become obscure.

HABITAT, DISTRIBUTION, and STATUS. Mainly a snake of the plains but in the Western Himalayas, recorded up to 1620 m at Almora. Peninsular India, Nepal, Pakistan, Sri Lanka. In the east up to the Teesta river. Not uncommon.

HABITS. An active little snake, mainly diurnal and seen most often during the rainy season. Not uncommon in the vicinity of human habitation where it makes its home in masonry of buildings and old walls. It climbs with facility. When alarmed, inflates its body to a remarkable degree and some specimens also flatten the posterior part of the head making the head more apparent than when normal.

FOOD. As with other species of the genus it feeds largely on reptile eggs but also takes mice and other mammals.

BREEDING. Gravid females have been obtained in August and young in April. It is believed that couples stay together, long after mating. The eggs number four or five and are remarkably elongate measuring 36 × 10 mm. Young apparently grow 10 to 12 cm in the first year attaining a length of 25 to 30 cm and 7 to 10 cm in the second year when they are 38 to 45 cm in length. At the end of third year they are over 50 cm in length. The smallest gravid female measured 55 cm.

RUSSELL'S OR VARIEGATED KUKRI SNAKE

Oligodon taeniolatus (Jerdon)

HARMLESS

SIZE. Largest measured 58 cm. Specimens exceeding 50 cm are uncommon.

IDENTIFICATION. The only snake found in the plains of the Indian peninsula having 15 scale rows throughout the length of the body and having the anal shield divided. A slender, graceful snake, with no indication of neck and uniform in girth. Tail short, about one-seventh total length.

COLORATION. Shows considerable differences in colour and markings. Five colour forms, all intergrading are recognised.

1. Buff, pale brown, cedar-brown, rarely light dun. Anterior lower margins of some of the scales conspicuously variegated by light and dark streaks. The dark streaks tending to congregate and form crossbars usually narrowly outlined with buff. Four longitudinal brown stripes, two broad, along the back, and two narrow, along the flanks. Stripes often obscure. A pale vertebral streak often interrupted. Belly unspotted. Found throughout its distributional range.
2. A median series of large well-defined, round spots, anterior ones often divided. Two rows of smaller spots flank the median series. No stripes. Variegation, if present, confined to flanks. DISTRIBUTION. India south of lat. 20 N.; Sri Lanka.

3. Dorsal spots larger than in 2 and longitudinally elongate. Edged with dark brown and twice as long as interspaces between them. 18 to 22 in number on body. DISTRIBUTION. Nilgiris; Madras.
4. Large; transversely arranged, dark brown, black-edged spots, 14 to 16 on body. DISTRIBUTION. Western Ghats; Madras District.
5. Large, dark brown, rounded spots, edged with black and white and paired or alternate on opposite sides of vertebral line. DISTRIBUTION. Kerala; Sri Lanka.

HABITAT, DISTRIBUTION, and STATUS. Indian subcontinent, from Baluchistan to Purnea, Bihar, Sri Lanka. Like *arnensis*, it is not uncommon near human habitations.

HABITS. Largely diurnal. A very common snake in the evergreen forests at higher elevations of the Western Ghats. Fond of basking in the sun on rock or grass. Feeds mainly on eggs of other reptiles and frog spawn. Little information is available on breeding habits. A female was seen in April with 9 eggs in a hole in a revetment. A young obtained in May measured 75 mm.

The majority of Kukri snakes are hill forms and a large number of species are found in the Western Ghats, below the Goa Gap and in the Eastern Himalayas and further east. A common species in the Eastern Himalayas is the LADDERBACK OF LIGHT-BARRED KUKRI SNAKE, *Oligodon albocinctus* distinguished by the 19 costals at mid body, entire anal and a single temporal. The coloration is brown of various shades, often strongly tinged purplish or red. Rarely uniform lobster or salmon red. Back with regularly arranged, black bordered, white, greyish white, grey or pale yellow, crossbars. 17 to 25 on the body and 4 to 8 on the tail. The head, pale grey or yellowish bears the characteristic three marks of Kukri snakes which may distintegrate and become confluent. Most of the specimens obtained have been from tea gardens in Darjeeling Dt., W. Bengal and Assam. It is also known from Chittagong, Bangladesh and in Burma as far south as Arakan Hills. Common in the eastern Himalayas. Adults average 60 to 76 cm in length.

Russell's or Variegated Kukri Snake (*Oligodon taeniolatus*) (See p. 83)

Common Indian Bronzeback or Tree Snake (*Dendrelaphis tristis*) (See p. 85)

COMMON INDIAN BRONZEBACK OR TREE SNAKE

Dendrelaphis tristis (Daudin)

HARMLESS

LOCAL NAMES. Hindi (Deccan) *Goobra*; Bengali *Bet anchora*; Marathi *Rookai, Maniar*; Gujarati *Mancas, Kudako, Oodto*; Tamil *Komberi moorken, Panaiyeri pamboo, Chitooriki pamboo*; Malayalam (Malabar) *Villooni*; Singhalese *Haldanda.*

SIZE. Largest measured ♂ 1.37 m, ♀ 1.69 (Smith).

IDENTIFICATION. Distinguished by the enlarged vertebrals and ridged ventrals; vertebrals narrowly enlarged obviously longer than broad. 15 costals at mid body. 2 labials touch eye. An elegant snake, with elongate head, bluntly rounded snout, small nostril and large lustrous eyes with golden iris and round pupil. Neck distinctly constricted, body long, slender, smooth and rather depressed (flattened from above downwards). Belly conspicuously ridged on either side. Tail long and tapering and nearly one-third the total length. Ridged below as in the belly.

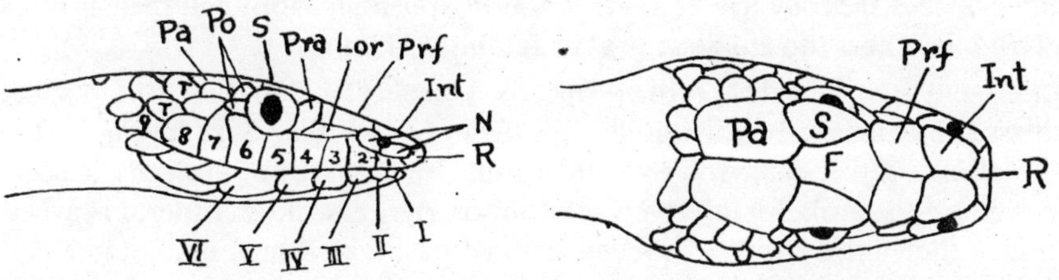

Head Scalation *Dendrelaphis tristis* For key to abbreviations, see Glossary

COLORATION. Dorsally uniform purple-brown, bronze-brown or rarely ruddy brown, except for the vertebral region which is more or less distinctly lighter and last row and a half of scales on the flanks which are yellowish. Vertebral stripe involves the vertebral and half the next row, conspicuous the whole length of the body or only anteriorly. Neck and forebody usually with a series of black streaks often paired and disjointed. A yellow, black-bordered flank stripe from neck to vent. Upper lip yellow, creamy-buff, or opalescent, abruptly demarcated above. A roundish yellow spot on parietal suture. 2nd to 4th supralabials (upper lip scales) black edged, often the 1st also. An obscure black post-ocular (eye) streak. Lower half of each body scale with a black edged light blue patch visible only when the snake is excited. Belly uniform creamy yellow, pale grey, greenish, or bluish green.

HABITAT, DISTRIBUTION, and STATUS. Bronzebacks live almost entirely on trees and shrubs, rarely coming to the ground. Peninsular India and Sri Lanka. Common in south India and along the Western Ghats. Uncommon north of

the Tapti in the Gangetic plain. Common in the Himalayan foothills.
HABITS. An active snake, of restless habit and quick movement. On ground it moves rapidly with forebody erect and ascends trees with truly amazing speed. The coloration is decidedly protective and when the snake is still, it bears a remarkable resemblance to a small dry branch. If detected its rapidity of movement and tenacious grip make it difficult to capture. It has also the ability to spring. A lively and plucky snake but disposition varies with individual temperament, some may be placid and others may strike when handled. When annoyed the body is expanded bringing into view the sky-blue marking of the scales. The neck is compressed and throat pouched before lunging forward to strike. Becomes quite tame in captivity.

A diurnal snake, it feeds largely on lizards and tree frogs; also snakes are taken, as well as nestlings of birds.

Oviparous; gravid specimens have been obtained from September to February and August. The period of gestation is believed to be four to six months and eggs hatch four to six weeks after laying. The eggs six to seven in number are remarkably long and measure 29-39 × 10-12 mm. Development commences before they are voided. Smallest specimen obtained measured 261 mm (10½ in.) and the smallest gravid female 521 mm (3ft 4 in.).

MISCELLANEOUS. In south India, there is a general belief that this perfectly harmless snake is fatally poisonous. It is believed that the snake having bitten a person climbs a tree near the pyre to watch the cremation and only descends after seeing the smoke rising from the funeral pyre. A mock funeral is arranged to save the victim for it is believed that as the snake thus duped, descends to the ground, the venom leaves the body of the victim, who recovers. Another strange belief in the Malabar district of Kerala is that the snake can thrust its tail into the ground, balance there on and assume the shape of a bow, hence the local name 'villoonie' derived from 'villoo', a bow.

PAINTED BRONZEBACK
Dendrelaphis pictus (Gmelin)

HARMLESS

Occurs in Bengal, Eastern Himalayas and countries to the east. Closely resembles *D. tristis* and can be separated with certainty only by the characters of maxillary teeth, the posterior maxillary teeth being longer than the rest as opposed to the posterior maxillary teeth being shorter than others in *D. tristis*. Bronze-brown above with a thin black stripe demarcating it, a yellow or cream flank stripe bordered below by a conspicuous black stripe almost as broad; commencing from behind the eye costal scales blue tipped as in *tristis*. Belly uniformly yellow, greyish of greenish. A bronze-olive or greenish, occasionally reddish race *D. p. andamanensis* occurs in the Andamans. In its habits the Painted Bronzeback is similar to the Common Indian Bronzeback.

Golden Tree Snake or Flying Snake (*Chrysopelia ornata*) (See p. 87)

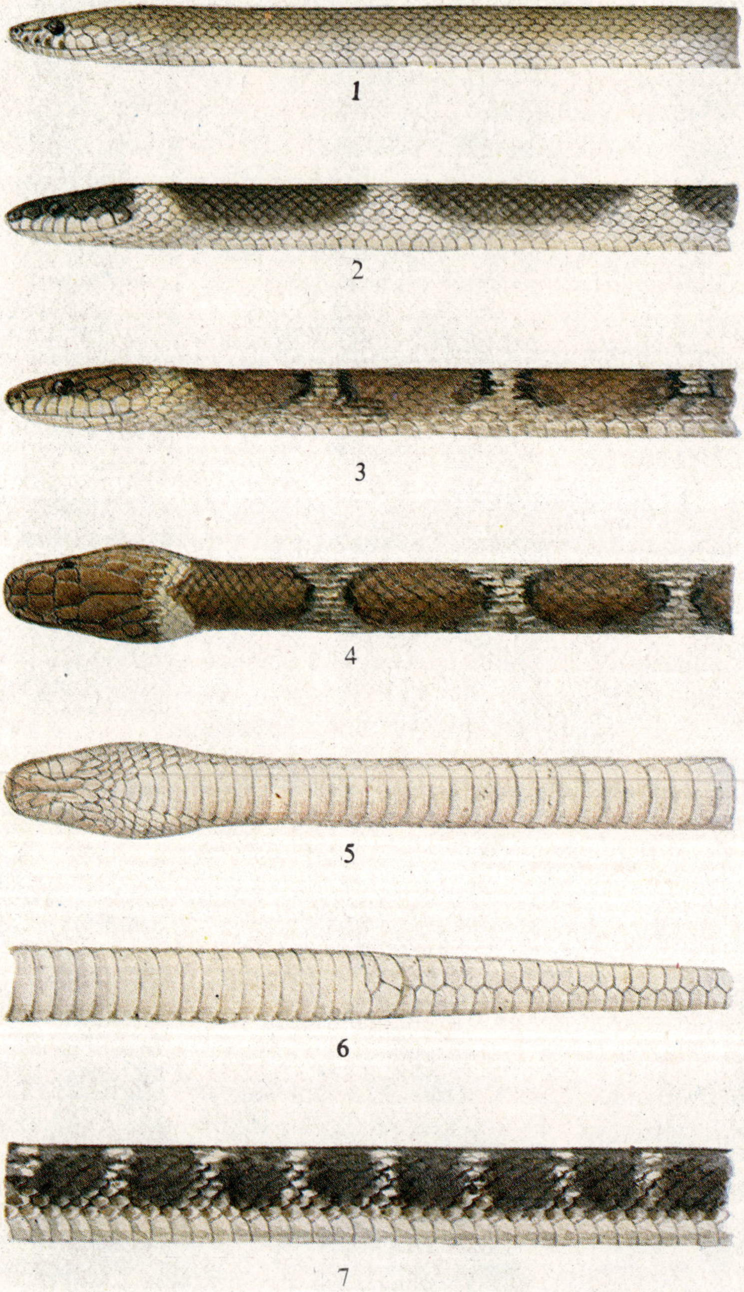

1-6. Common Wolf Snake (*Lycodon aulicus*); 7. Shaw's Wolf Snake (*Lycodon striatus*) (See pp. 89, 90)

GOLDEN TREE OR FLYING SNAKE

Chrysopelea ornata (Shaw)

HARMLESS

LOCAL NAMES. Urdu *Kalajin*; Singhalese *Pol-mal-karawala, Mal karawala.*

SIZE. Largest measured 1360 mm (4 ft 5½ in.). According to Smith, specimens measuring 1400 mm (4 ft 7 in.) are not uncommon.

IDENTIFICATION. The lateral, sharp and pronounced keeled condition of the ventrals in association with the un-enlarged vertebral row of scales distinguish this snake. The coloration is also distinctive. The body though slender is far less so than in other tree snakes. Head pear-shaped strongly depressed. Eyes large with round pupil. Neck distinctly constricted. Belly with sharp lateral keels. Tail about one-fourth total length.

COLORATION. Young, black above with narrow pale greenish yellow crossbars, dilated or not, vertebrally and on sides. Scales with or without a dark mesial streak. With age the extent of green increases.

Two major colour forms known in the adult. In Sri Lanka and Western Ghats below Goa, greenish yellow or pale green with each scale having a black mesial streak or spot and more or less edged with black and with completely dark black crossbars at intervals. A series of large, flower shaped, reddish or orange vertebral spots present or absent. Ventrals greenish, outside the keel edged with black or spotted. Head black with yellow crossbars and spots. The flower-shaped spots commoner in Sri Lanka than in south India. The second colour variety occurring in Burma and Siam lacks the vertebral spots and the black crossbars are much less evident.

HABITAT, DISTRIBUTION, and STATUS. Forests of the Andamans, Western Ghats, south of Goa Gap, Sri Lanka, North Bihar, West Bengal eastwards throughout the Indochinese subregion. A common snake in Indochina and Sri Lanka, it is rarely seen in its Indian range.

HABITS. This handsome snake is essentially arboreal but is also frequently seen in grass and on low bushes. A decidedly plucky and occasionally fierce snake but individual temperaments vary. Diurnal, it feeds largely on geckos and other lizards (*Draco*). Bats are known to be taken. A graceful and agile climber, moving from branch to branch or up a perpendicular tree trunk with consumate ease. Several instances are available of their ability to spring horizontally and upwards and to glide from a height to the ground or to another tree. In gliding, the snake after launching, straightens itself rigidly and hollows the belly between the ridges on the ventrals, thus producing a mechanical effect impeding the action of gravity, buoying up the body and retarding the speed of descent. The distance covered is often considerable. An instance is recorded of this species gliding a distance of about 50 m when first seen it was at a height of about 6 m and had apparently launched itself from a tree higher up the slope. It passed the observer at chest level and at the end of the glide landed on the ground near some bushes. It is said to have glided "with a furious swimming motion keeping itself perfectly horizontal". The habit of clinging head downwards with body in a wide 'S' on the bare, perpendicular trunk of trees is perhaps preparatory to launching itself into a glide.

Very little information is available on breeding habits. Gravid females have been obtained in May and June and hatchlings in June. According to Smith pairing takes place in June in Bangkok. Six to twelve elongate eggs are laid. Hatchlings measure 114-152 mm (4½ to 6 in.) in length. Smallest gravid female recorded was 1093 mm (3 ft 7 in.) in length.

COMMON WOLF SNAKE

Lycodon aulicus (Linnaeus)

HARMLESS

LOCAL NAMES. Hindi *Garar* (U.P.); Bengali *Kaurialla*; Gujarati *Chitalun*; Marathi *Kandya*; Malayalam *Shunguvirian*; Tamil *Valappanayan, Kattu virian, Sanku virian, Vellikel viriyan*; Singhalese *Tel karawala, Alu polonga.* Scientific Lycodon = wolf toothed, Aulicus = house dweller.

SIZE. Grows to about 765 (2½ ft) in length but specimens over 610 (2 ft) are rare.

IDENTIFICATION. Essential characters for identification are; 17 scale rows at mid body, a loreal touching internasal, nasal touching 1st and 2nd labial and presence of 9 supralabials. A glossy, slender snake with a pear-shaped depressed head. Eye black, tongue pinkish with white tip. Neck slightly constricted. Tail about one-fifth to one-sixth the total length.

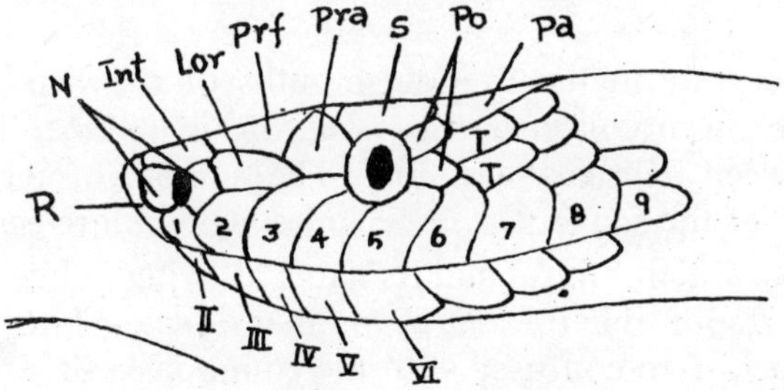

Head Scalation *Lycodon aulicus* (× 2)

Key Characters: Loreal (Lor) touching internasal (Int), Nasal (N) touching 1st and 2nd supralabial (upper lip shield), 9 supralabials

COLORATION. Colour and markings variable. Two races are recognised. In race *Lycodon aulicus aulicus* the ground colour is light brown to dark cigar brown with yellow crossbars on body and a collar of the same colour or two whitish blotches on occiput. Bars expand on the flanks to enclose islets of the ground colour. Bars may be present the whole length of the body or confined to the anterior region of the body. Bars 9 to 18 (12-19 Smith) in number.

In *Lycodon aulicus capucinus* the ground colour is dark brown or purplish brown with distinct white or yellow reticulations.

In both races occasionally specimens without any markings and uniform brown in colour occur.

HABITAT, DISTRIBUTION, and STATUS. The snake most often seen near and in human habitations. Sri Lanka, Indian subcontinent, Maldive Islands. Burma to Indochina, south China, Malaya, Indonesia, Philippines. One of the commonest snakes of the Indian plains.

HABITS. Among snakes, the wolf snake is the one that seems to seek out and profit by a human environment. The common wolf snake is the species most often met within houses whether in the country or heavily congested urban areas. It hides during the day in crevices in masonry or beneath boxes, stones or any other convenient hideout. A lively snake, it readily strikes when provoked. Like most other snakes, however, it soon gets accustomed to being handled in captivity. If in the open and baulked in its endeavours to escape, it will frequently coil itself into a heap and if worried will hide its head beneath the coils. Often, while lying thus it fixes its coils rigidly so that it can be tossed into the air without it releasing its folds as one might do a piece of knotted cane.

An excellent climber, capable of going up almost smooth vertical surfaces, climbing with the aid of its ribs and the free borders of its belly shields. The habit of climbing and living in between the ceiling and the roof often leads to its falling amidst the unwary family circle. The wolf snake prefers lizards of the Gecko family but takes any small animal it can overcome. Mice and skinks are also eaten.

BREEDING is apparently in the first six months of the year when several gravid females have been collected. Eggs varying in number from four to eleven are laid between February and July. The number of eggs in a clutch depends on the size of the snake, the larger ones laying more eggs.

The readiness with which it bites and its habit of living in houses make this snake undoubtedly responsible for a large number of snake bite cases in India every year. It is most often confused with the common krait *from its almost identical colour pattern and the long teeth on the upper and lower jaws which are mistaken for fangs*. Thus misidentified it helps to bolster the reputation of the numerous antidotes to snake poison which get the credit of having averted death. Any fatality resulting from the bite of the wolf snake is purely the result of fright.

Period of gestation and incubation unknown. Young at birth are 17 to 19 cm in length, and growth doubles this length in the first year. Sexual maturity is attained at about two years. Smallest gravid female noted was 45 cm in length.

SHAW'S WOLF SNAKE

Lycodon striatus (Shaw)

HARMLESS

Resembles very closely the Common Wolf Snake in body characters and habits. A small snake (430 mm) it is distinguished from *aulicus* by the presence of 8 instead of 9 supralabials and lesser number of ventrals 154-195 and caudals 35 to 58. Dark brown or black above with 11 to 18 white or yellowish

Checkered Keelback (*Xenochrophis piscator*) (See p. 92)

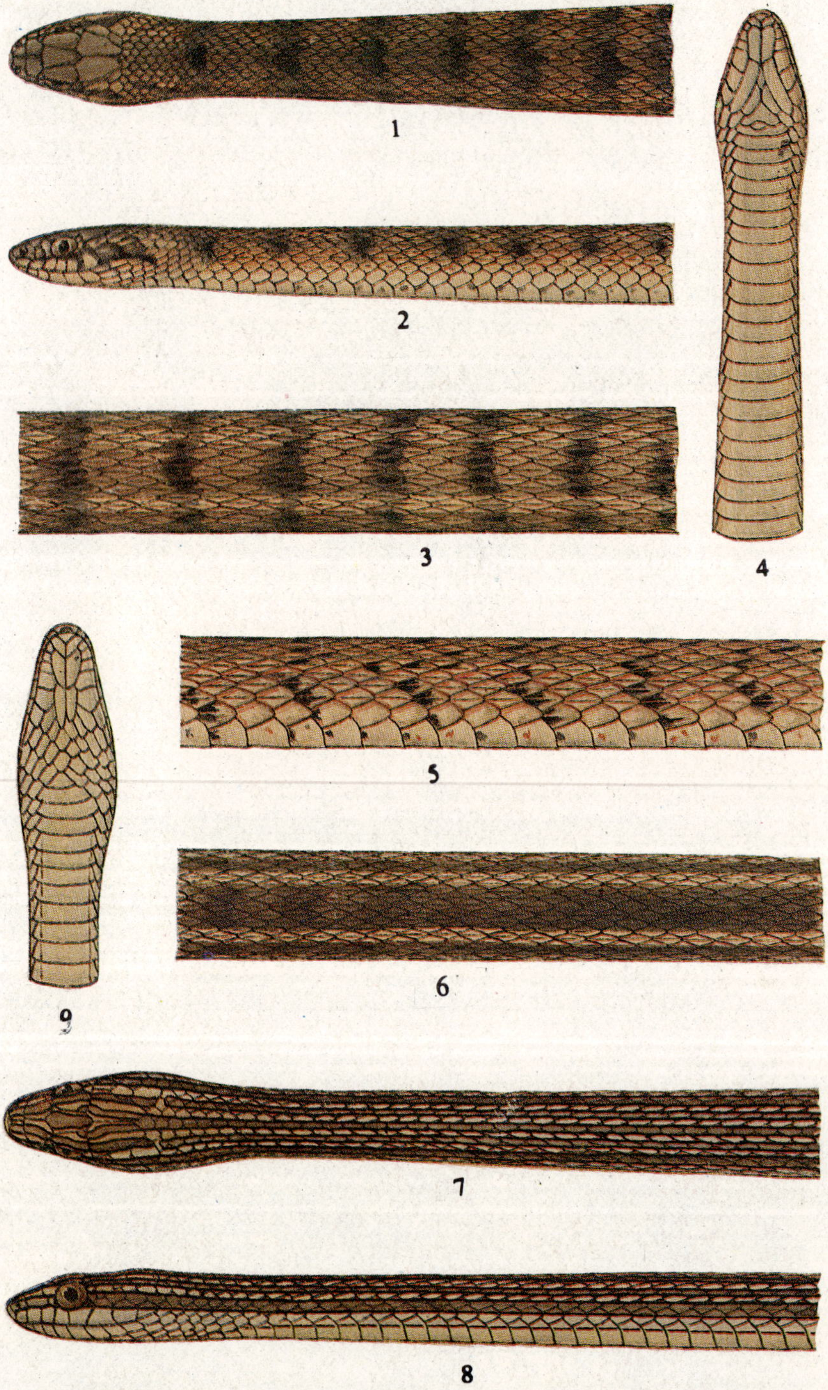

1-6. Buffstriped Keelback (*Amphiesma stolata*); 7-9. Condanarous Sandsnake (*Psammophis condanarus*) (See pp. 94, 98)

crossbars on body which divide on the sides to enclose triangular spots of the ground colour. Belly and upper lip white. A timid snake which hides its head beneath its coils if disturbed. Sexes are found together in August when the female is egg bound or has laid. Eggs 2 to 4 in number.

DISTRIBUTION. Sri Lanka, peninsular India, westwards to Iran and to the east up to Chota Nagpur.

TRAVANCORE WOLF SNAKE
Lycodon travancoricus (Beddome)

HARMLESS

Another common species of the hill areas of southern India, the Travancore Wolf Snake (*Lycodon travancoricus*) is distinguished by the loreal scarcely touching the internasals, 9 supralabials, ventrals 176-206, caudal 64-76 (paired or sometimes some or rarely all single) and single anal. In colour it is purplish brown or blackish above with pale yellow crossbars.

CHECKERED KEELBACK

Xenochrophis piscator (Schneider)

HARMLESS

LOCAL NAMES. Hindi *Pani ka samp, Dhoria*; Oriya *Pani dhanda*; Assamese *Dhora*; Gujarati *Dendu*; Bengali *Jal dhonra*; Marathi *Virola, Pan chidda, Divad*; Kanarese *Neeru havu, Holay havu, Haramandalatha havu*; Malayalam *Neer kolee, Neer mandalee*; Tamil *Thaneer pamboo*.

SIZE. Longest measured 1.48 m (4 ft 10 in.). Normally rarely exceeds 1.21 m.

IDENTIFICATION. Distinguished by a combination of characters which are 19 costals at mid body, 2 (4th and 5th), supralabials (upper lip scales) touching eye, presence of pair of internasals and an undivided anal shield. A fairly robust snake with oval head having slit-like nostrils and moderately large eyes both with a decidedly upward inclination. Tail one-third to one-fourth total length.

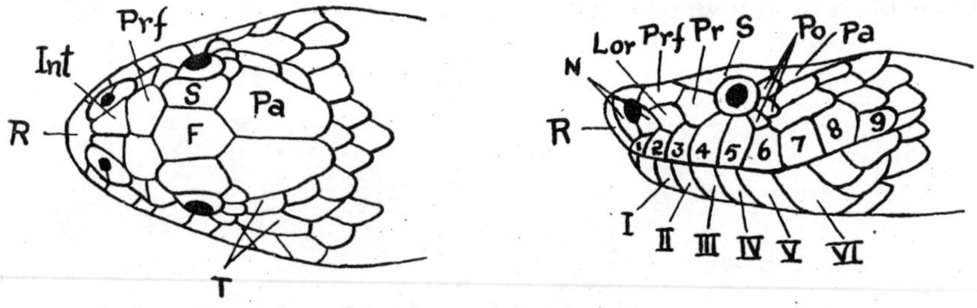

Head Scalation *Xenochrophis piscator* (nat. size)

Key Characters: 4th, 5th supralabials (upper lip scales) touching eyes. A pair of internasal shields (Int.)

COLORATION. Variable, four races are recognised. *Xenochrophis piscator piscator* with strongly keeled scales and five rows of black spots on an yellowish or olivaceous background, spots varying in size sometimes occupying most of the back. Number of rows not constant throughout the body. Head olive-brown with two black streaks one below and one behind eye. Belly white or yellowish. This race is found throughout the Indian subcontinent from Baluchistan to Assam into upper Burma. A pale form occurs in the Himalayas. Hatchlings may have two light spots on the parietals which disappear with age.

Generally the ground colour may be dull green, olive-brown or brown of almost any shade, light or dark. Apart from the black markings some are speckled, spotted or blotched with red varying in intensity from salmon or rose pink to a brillant scarlet. This lively ornamentation is almost entirely confined to basal half of scales, seen to best advantage when the snake dilates itself under excitement.

1

2

3

4

5

6

1-3. Schneider's Smooth Water Snake (*Enhydris enhydris*);
4-6. Olivaceous Keelback (*Atretium schistosum*) (See pp. 104, 96)

1

2

3

4

5

Green Keelback (*Macropisthodon plumbicolor*)
1-3. Adult; 4-5 Juvenile (See p. 97)

HABITAT, DISTRIBUTION, and STATUS. The commonest freshwater snake. In addition to areas listed under the races, it occurs in the Malayan region, south China and Taiwan.
HABITS. One of the commonest of Indian snakes. Frequents water and is very common in tanks, paddy fields, pools and rivers. In swampy area, may be seen away from water. Wall considered this snake to be the most vicious among Indian snakes with the exception of the *Echis*. Strikes rapidly and with great determination holding on tenaciously. Erects and flattens forebody prior to striking. But temperaments vary and some are very mild. An extremely active snake, capable of jumping clear off the ground and will do so repeatedly if pursued. In water it swims nimbly and with vigour, and is a versatile diver. In arid areas the snake aestivates during the summer appearing again with the monsoon. In the north it hibernates during the winter but on warm days individuals are often seen sunning themselves on vegetation above water level.
FOOD. Feeds mainly on frogs. Fish are frequently taken and also tadpoles. No attempt is made to constrict and frogs may continue to squeal for sometime after being swallowed.

One snake was noticed lying on top of an outlet pipe from a lake trying to catch fish coming out through the pipe (62). Another was seen in a water hole catching tadpoles. The snake usually faces upstream when seen in small streams.
BREEDING. The female is longer in total length but has a shorter tail. Pairs apparently stay together for a considerable time, even after the female is egg bound. Copulation has been noticed from October and females with eggs have been obtained from November to May. The period of gestation is 55 to 67 days. One of the most prolific of Indian snakes, clutch size varies from 8 to 91 eggs. The clutch of 91 eggs was voided over a period of 12 hours. Eggs, 18 × 27 mm in size. Period of incubation varies from 37 to 51 days. The female probably incubates the eggs. The eggs are usually laid in a nest hole in the ground near water. Length of hatchling varies from 170 to 215 mm and young apparently gain about 254 mm in a year averaging 431 to 457 mm in length in the 1st year, 685 to 711 mm the 2nd year and 939 to 965 mm the 3rd year. The smallest gravid female measured 806 mm (2 ft 7 in.).
MISCELLANEOUS. A curious behaviour of this snake is the habit of shamming death when attacked by a predator like the mongoose. When its strikes are ineffective it sinks to the ground and in slow motion turns over on its back maintaining a slow muscular movement like a snake with its head beaten in. When the danger is past, the snake will beat a hasty retreat but will repeat the performance if caught again (51).

The secretion of the large parotid glands is fatal to small mammals, birds, lizards and frogs in laboratory experiments, but has no effect on man and larger mammals (30).

BUFFSTRIPED KEELBACK
Amphiesma stolata (Linn.)

HARMLESS

LOCAL NAMES. Bengali *Hele sáp*; Marathi *Naneti*; Telugu *Wannapam*; Tamil *Nirkatan pambu*.

SIZE. Longest measured 900 mm (M.A. Smith). Females are consistently longer than males which rarely reach 620 mm (2 ft) in length.

IDENTIFICATION. Distinguished by the nasals not touching the 2nd supralabial, rostral touching 6 shields, a single anterior temporal and 19 rows of costals. The buff stripes along the sides are also distinctive. A graceful little snake with a relatively short body and long tail which is approximately a quarter of the total length. Neck fairly evident. Body tapers to vent. Eyes large, pupil round, flecked with gold. Tongue, dull orange, black tipped.

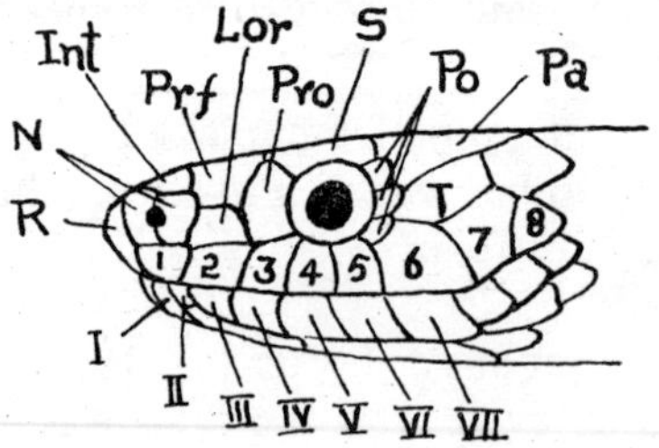

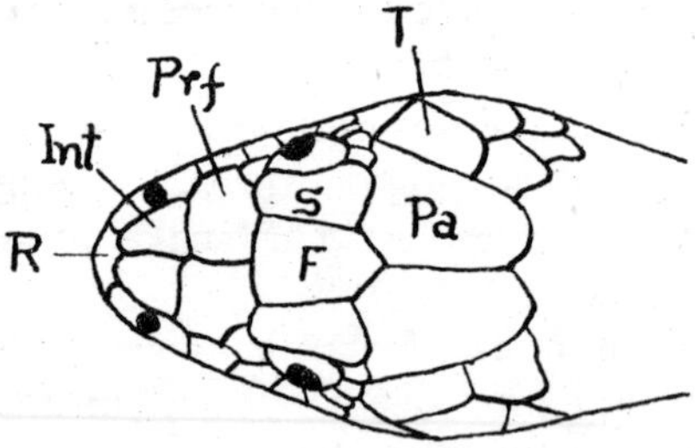

Head Scalation *Amphiesma stolata* (×2)

Key Characters: Nasal (N) not touching 2nd supralabial (upperlip shield). Rostral (snout shield) touching six shields. A single anterior temporal (T) shield

COLORATION. Ground colour olivaceous-brown of varying intensity. A pair of conspicuous buff stripes covering one whole or two half rows of scales front neck or forebody to tip of tail. Some black or blackish, irregular crossbars on body, olivaceous anteriorly. Head olivaceous-brown, whitish, yellowish or orange on lips. Supralabials margined posteriorly with black. Chin and throat white, yellow or orange. Belly white with some small scattered black spots.

Two distinct colour varieties are known. In the typical form the overlapping margins of scales are blue-grey or pale blue. In the variety *erythrostictus* bright vermillion replaces the blue. These colours are very evident when the snake inflates its body under excitement. The typical variety is found throughout the distribution but *erythrostictus* is more common in coastal areas.

HABITAT, DISTRIBUTION, and STATUS. Inhabits banks of rivers etc. and marshy areas, from Sind in the west to south China in the east; Indochina. In India it is a common snake in the plains and rarer in the hills.

HABITS. A remarkably inoffensive and gentle little snake, common in fields, grassy and cultivated areas of open country during the rains, its choice of habitat being related to its food: frogs and toads. Often hides in holes in the ground or in masonry of drains, culverts. Essentially diurnal, when alarmed, some flatten the neck and forebody and distend themselves by deep inhalations bringing into view the beautiful blue or vermillion on the base of the scales. Aestivates during the hot weather, appearing abundantly only during the rains. Hibernates during the cold weather in northern India in soil at depths of 228 to 457 mm among grass roots.

FOOD. Its food is almost entirely frogs and toads. Indian specimens feed largely on the skipper frog (*Rana cyanophlyctis*), young bull frogs (*Rana tigerina*) and young common toads (*Bufo melanostictus*). Specimens in captivity have taken 131 toads in a year.

BREEDING. Sexes are evenly balanced in most parts of the country but in Assam females appear to be twice as numerous. Mating apparently occurs during aestivation and gravid females have been obtained from April to August and eggs are laid from May to September, in any convenient refuge underground. Females often remain with the eggs. Incubation is believed to vary with temperature being longer in the hills. The shortest period recorded is a month. Eggs which are pure white in colour, adhere and form a cluster when laid. Eggs measure from 22 × 16 to 35 × 18 mm. Clutches vary in number from one to fourteen but five to ten are the usual number.

Hatchlings measure 133 to 177 mm, females being slightly longer and the difference is maintained and becomes increasingly apparent with age, the difference in length exceeding 50 mm or more in favour of the female by the second year. The young are believed to double their length by the first year of life and treble it by the second year. The smallest gravid female measured 442 mm and many were 457 to 470 mm the length attained at the end of the second year of life.

Olivaceous Keelback
Atretium schistosum (Daudin)

HARMLESS

LOCAL NAMES. Telugu *Nalla wahlagillee pam*; Kanarese *Barmmya.*

SIZE. Longest measured 870 mm. Females are longer averaging 73 cm and males 55 cm.

IDENTIFICATION. Distinguished by a combination of characters, namely a single internasal, 19 costals at mid body and 8 or 9 supralabials. A fairly robust small snake with a short snout and small slit-like nostrils placed rather high as in water snakes. Body rough from the keeled scales. Tail long, one-fourth to one-third total length.

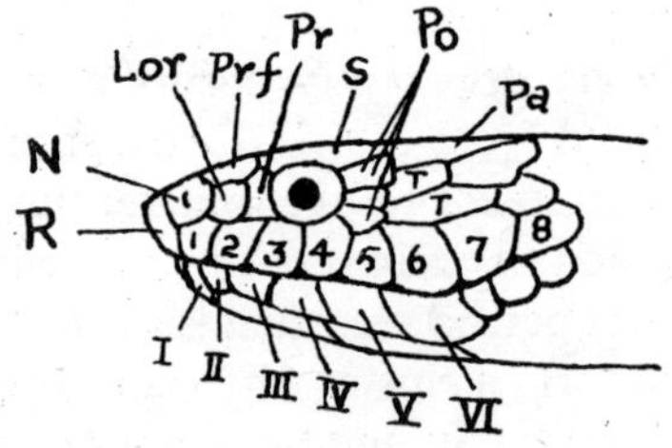

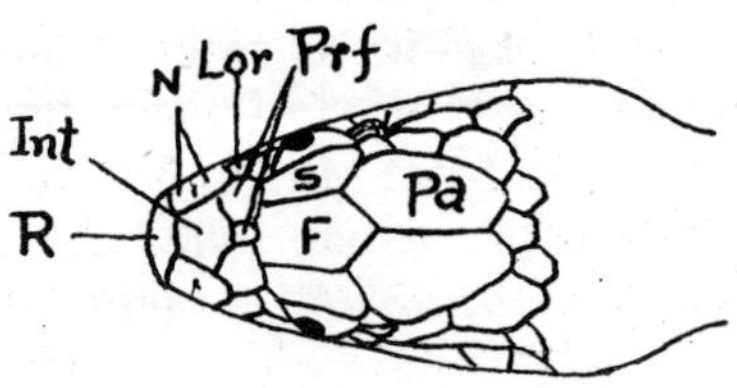

Head Scalation *Atretium schistosum* (nat. size)

Key characters: A single internasal (Int): 8 or 9 Supralabials (upper lip shields).

For key to abbreviations, see Glossary

COLORATION. Uniform deep olive-green above and uniform yellow of varying intensity below, sometimes tinged with pink or lilac on the flanks. The two colours abruptly demarcated on the penultimate row of costals. Lips yellow or pinkish. Specimens from south India with a reddish line along the 5th and 6th or 4th and 5th row of scales up to the vent. The line is brighter in the males.

HABITAT, DISTRIBUTION, and STATUS. Usually near water. Peninsular India south of 15°N. latitude and along east coast to Uttar Pradesh. Sri Lanka.

The snake is very common around Bangalore and also in North Arcot District of Tamil Nadu; and the Coconada area of Andhra Pradesh.

HABITS. The olivaceous keelback lives either in water or among the surrounding vegetation. Largely diurnal and feeds mainly on frogs, fish and crabs. Aestivates in hot weather.

BREEDING. Breeds during the rains and eggs are laid from January to April. The eggs 10 to 32 in number are white, soft and 30 to 35 mm in length. Hatchlings measure 167 to 174 mm. The smallest gravid female measured 717 mm.

GREEN KEELBACK

Macropisthodon plumbicolor (Cantor)

HARMLESS

LOCAL NAMES. Tamil *Pachai nagam*; Marathi *Hara samp, Gavathia, Gautya sap*.

SIZE. Longest measured 940 mm; longest male 750 mm.

IDENTIFICATION. Frontal shield in contact with 6 shields, 17 or 19 rows of costals two head lengths in front of vent. The *shields on the head* distinguish it immediately from the green pit-vipers which have none and the *round pupil* from the green whipsnakes. A small, rather stout, snake. Eyes moderately large, iris greenish gold. Tail short, one-seventh to one-ninth total length.

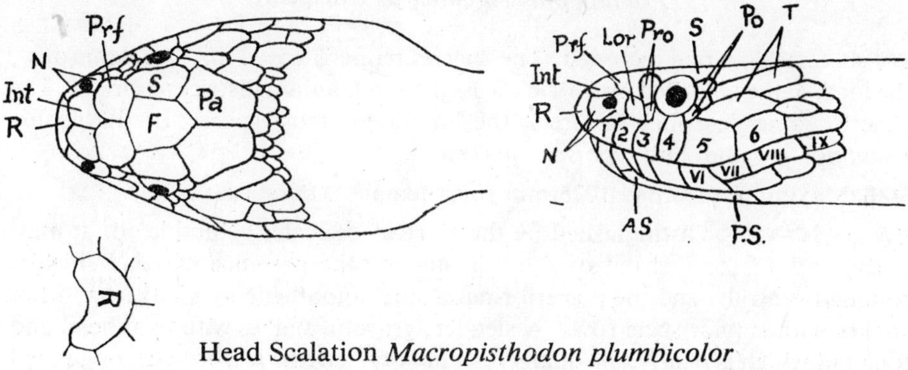

Head Scalation *Macropisthodon plumbicolor*

Key Character: Frontal (snout) shield (R) in contact with six shields

For key to abbreviations, see Glossary

COLORATION. Adults, uniform grass-green above with a few black spots which may show a tendency to form transverse bars. A few small white spots on body; upper lip, chin, throat and belly uniform white or belly may be greenish or plumbeous. Young have a well-defined black chevron on the nape with its angle pointed forwards. Following the chevron is a broad gorget of bright yellow or orange. A black streak from eye to gape. The green colour in this species is produced by a yellow pigment overlying the bluish scales.

HABITAT, DISTRIBUTION, and STATUS. A hill species. Occurs throughout India except the Ganges valley and the extreme northwest Sri Lanka.

HABITS. Mainly a snake of the hills favouring altitudes between 600-1800 m. Frequents grass and low vegetation. A singularly gentle and inoffensive snake, it erects its forebody and flattens its neck like a cobra, hence the Tamil name 'green cobra'. Though often seen in grass and scrub during daytime it restricts its movements to the night. Aestivates during the summer.

FOOD. Feeds more or less exclusively on the Common Toad (*Bufo melanostictus*) but occasionally takes other amphibians and lizards. The secretion of the large parotid glands is toxic to frogs and toads (30).

BREEDING. Breeding habits not well known, eggs have been obtained in March-June and hatchlings from April to September. Eggs 7 to 16 in number. Hatchlings measure 136 to 168 mm and smallest gravid female 450 mm (1 ft 5¾ in.) suggesting one and a half to two years of age.

CONDANAROUS SANDSNAKE

Psammophis condanarus (Merrem)

HARMLESS

LOCAL NAMES. None recorded. The species name is based on 'Condanarous', the local name recorded by Patrick Russel from Ganjam district of Orissa. The generic name describes in Greek the habitat of most species of the genus: *Psammos* = sand, and *Ophis* = snake.

SIZE. Maximum recorded 1075 mm for a female. Males are smaller.

IDENTIFICATION. Distinguished by the 17 rows of scales behind head, at mid-body, and 13 rows ahead of the tail, and by the presence of head shields, rounded ventrals, and the posterior nasal shield not being divided; from other snakes with similar scale rows. A slender, graceful snake, with oval head and long tail which is nearly one quarter of the total length. Iris brown, tongue red with black tip.

COLORATION. Dorsally streaked alternately with distinct nut brown and greenish olive or buff stripes. Belly sulphur or primrose yellow. Skin between scales blackish.

HABITAT, DISTRIBUTION, AND STATUS. Arid and semi-arid country in Pakistan and peninsular India up to Bengal in the east and Andhra Pradesh in the south. Also in light forested country in western Himalayas; Burma. Not uncommon in some parts of its total distribution, seven being obtained from the crop of a Short-toed Eagle (*Circaetus gallicus*) shot in the Dun. It is one of the poisonous snakes with fangs in the back of the upper jaws.

HABITS. A diurnal snake of grassland and open jungle and is not as much a desert form as its allied species. Feeds on lizards, frogs and occasionally other snakes. One specimen had swallowed a saw-scaled viper.

Precise information on breeding is not available but it is believed to breed in the summer.

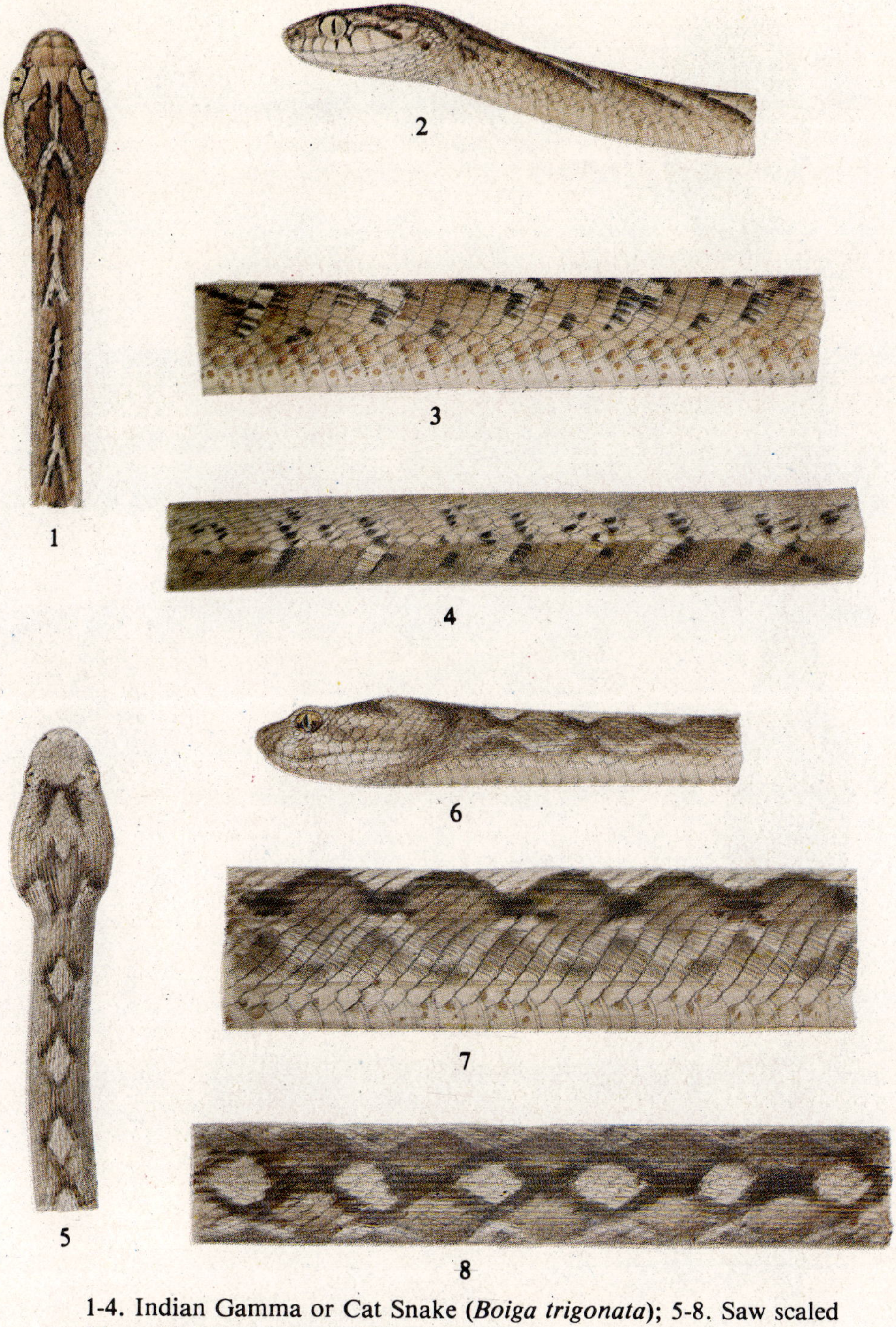

1-4. Indian Gamma or Cat Snake (*Boiga trigonata*); 5-8. Saw scaled Viper (*Echis carinatus*) (See pp. 94, 124)

1-4. Green or Bamboo Pit Viper (*Trimeresurus gramineus*);
5-9. Common Green Whip Snake (*Ahaetulla nasutus*) (See pp. 126, 102)

INDIAN GAMMA OR CAT SNAKE
Boiga trigonata (Schneider)

HARMLESS

LOCAL NAMES. Gujarati *Kodiyo sap*; Marathi *Manjra*.

SIZE. Longest measured ♂ 965; ♀ 1170 mm (Smith).

IDENTIFICATION. Distinguished by a combination of characters, namely costals or body shields 21:21:15; preocular not reaching upper surface of head, vertebrals feebly enlarged and subcaudals 75 to 96. A small snake with slender, markedly, compressed body, neck distinctly constricted,' eyes large with mustard yellow iris and vertical pupil.

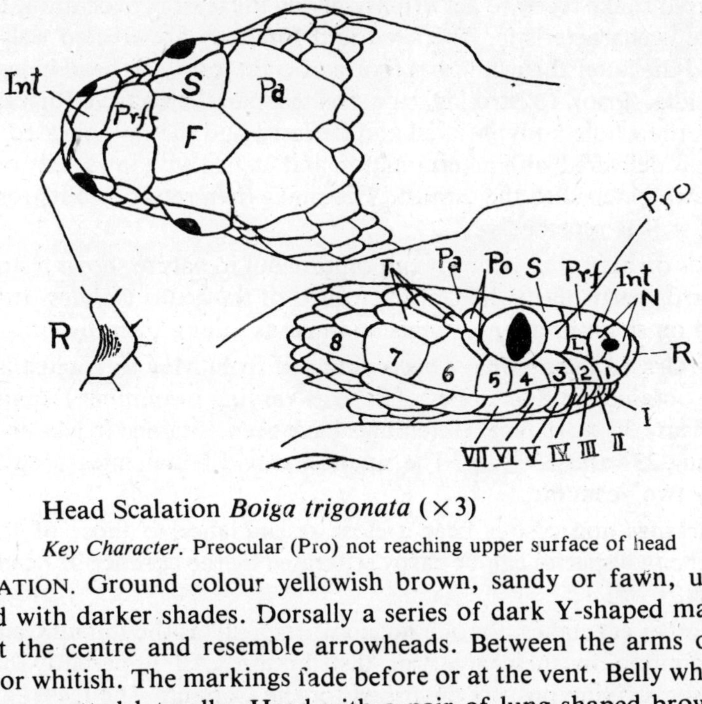

Head Scalation *Boiga trigonata* (×3)

Key Character. Preocular (Pro) not reaching upper surface of head

COLORATION. Ground colour yellowish brown, sandy or fawn, uniform or mottled with darker shades. Dorsally a series of dark Y-shaped marks which meet at the centre and resemble arrowheads. Between the arms of each Y, lighter or whitish. The markings fade before or at the vent. Belly white, rufous or brown-spotted laterally. Head with a pair of lung-shaped brown patches often bordered with black. A narrow dark streak from behind eye to gape.

The variety *melanocephala* (Annandale) from the Baluchistan border has a black head and is generally darker.

HABITAT, DISTRIBUTION, and STATUS. Essentially arboreal, frequenting bushes, scrub or trees, usually close to the ground. Inhabits areas without a shred of vegetation also. The whole of the Indian peninsula from Baluchistan (Pakistan) in the west to Assam in the east; Sri Lanka. A fairly common snake in India ascending up to about 1500 m in the western Himalayas.

HABITS. A nocturnal snake usually encountered on the move at night. In bushes it coils itself into a little heap and does not recline with body extended as other snakes do.

In south India, very common in evergreen and deciduous forest. An excellent climber it can jump from a good height to the ground. If picked up by the tail climbs up its own body and bites. Sometimes when caught by the neck ejects an evil smelling stream of yellow and white secretion from its anal gland, which has a smell difficult to remove (37). This snake and others of the genus have fangs in the back of the mouth, the last two teeth on the upper jaw being set apart from the rest in the form of grooved fangs. The secretion of the parotid gland is toxic and lethal to lizards and highly toxic to mice.

An intrepid snake ready to act offensively on the least provocation. Its striking posture is characteristic. The head and forebody are erected well off the ground and the latter thrown into a figure of eight loop, the head being poised in the middle. Prior to striking, the erected part is swayed forwards and backwards the whole body inflated and deflated and the tail vibrated briskly. The stroke is delivered with determination and malice with jaws fully open and the loops straightened to the utmost. The snake then retracts to its former attitude and strikes repeatedly.

FOOD. Feeds on almost anything it can capture but in nature shows a strong liking for lizards particularly the Garden lizards of the genus *Calotes*. In captivity, has fed on small birds, mammals and lizards killing them by constriction.

BREEDING. Gravid females have been obtained from May to August and eggs have been obtained in September. The eggs varying in number from three to eleven measure 30 × 10 mm. Hatchlings have been obtained in March and July measuring 237 and 260 mm. The smallest gravid female measured 570 mm apparently two years old.

The markings and colour bear a close resemblance to those of the *Echis* viper but the two species can be easily separated by the absence of head shields in the viper.

Several other catsnakes, though not so widespread as the Indian Gamma are equally common in their areas of distribution. All apparently have the characteristic striking posture described for the Gamma. The EASTERN GAMMA *Boiga gokool* (Gray) differs from *trigonata* in the strongly enlarged vertebrals; costals 21 (19): 21 (19):17; ventrals 219-232 and caudals 87-103. The pattern closely resembles that of *trigonata,* of which, it is apparently the eastern form occurring from Darjeeling eastwards.

The CEYLON CAT SNAKE *Boiga ceylonensis* (Günther) another closely related species occurs in addition to Sri Lanka in the Western Ghats, Eastern Ghats, Assam, and Andaman Islands. Distinguished by the pre-ocular extending to the upper surface of the head. Vertebrals strongly enlarged. Costals in 19, 21 or 23 rows, ventrals 214-267, caudals 90-133. Another common and fairly large species growing to over 2 m is FORSTEN'S CAT SNAKE. *Boiga forsteni* (Dum. & Bibr.) which occurs along the Western Ghats from Mt. Abu southwards, Eastern Ghats, Ganges valley and eastern Himalayas. In colour it is dirty whitish or buff, powdered or marbled with brown. A series of large brown crossbars are distinct anteriorly and disappear in the mottling behind. In its habits it resembles the Indian Gamma but is more arboreal and feeds on larger prey including bats, pigeons and poultry. Costals 25 or 27:27 or 29:17 rows, vertebrals feebly or strongly enlarged, variable even in the same species with a distinct lateral keel. Caudals 102-119. They are said to live in pairs and to frequent holes in Mohwa (*Madhuca latifolia*) trees in Orissa.

COMMON GREEN WHIP SNAKE
Ahaetulla nasutus (Lacépède)

HARMLESS

LOCAL NAMES. Hindi *Harahara*; Marathi *Sarpa toli, Haran tol*; Gujarati *Málana, Lile-jad sáp*; Bengali *Lan doga sap*; Tamil *Kankothi pamboo, Pachai pamboo*; Malayalam *Pachila pambu*; Telugu *Pasarika pamboo*; Singhalese *Ehetulla;* Burmese *Mywe sein*.

SIZE. Longest measured 1944 mm (6 ft 4 ½ in.) female. Longest male 1325 mm.

IDENTIFICATION. The green colour, the horizontal pupil and the pointed snout are distinctive. A long, very slender snake with elongated snout narrowing rapidly and terminating in a projecting dermal appendage. Eyes large, iris powdered with gold, pupil horizontal. The markedly concave loreal region permits binocular vision. Tongue pale pinkish with white tip. Mouth pale pinkish inside. Tail cylindrical, very long, being sometimes in males more than one-third total length, and longer relatively than in any other Indian snake.

COLORATION. Above bright verdant-green. Belly of an equally intense but lighter shade of green, adorned on each side by a well-defined, narrow, white (sometimes bluish) stripe situated on the ventral shields. Chin and throat white with light sky blue and yellow mottlings. Under excitement a striking black and white ornamentation of oblique lines is exposed on the anterior two-thirds of the body. This coloration is confined to the skin and lower edges of the scale. The green colour is produced by interaction of an yellow pigment on the scales with the blue of the skin. The blueness of the skin is itself a refractory colour like the blueness of the sky.

HABITAT, DISTRIBUTION, and STATUS. Peninsular India excluding Ganges valley west of Patna, eastwards to Burma, Sri Lanka. Essentially an inhabitant of the plains and low uplands but recorded up to 1800 m. Seen most frequently on low bushes and scrub in jungles and gardens and groves in populated areas. A common snake.

HABITS. It is diurnal and usually reclines on the topmost boughs of bushes and escapes notice by its cryptic coloration. An elegant snake with a wonderful turn of speed over foliage, the slender and light body is capable of obtaining support from the minutest twig and twining stem.

Usually a gentle snake it could be very fierce when freshly caught. When alarmed it rears its head and forebody and the dilation of the forebody brings into prominence the black and white chequering of the skin. If further excited the jaws are opened widely and the lower jaw expanded laterally by separation of the mandibles by strong muscular action so that the lower jaw becomes shovel shaped and considerably increased in area. The lower lip is turned

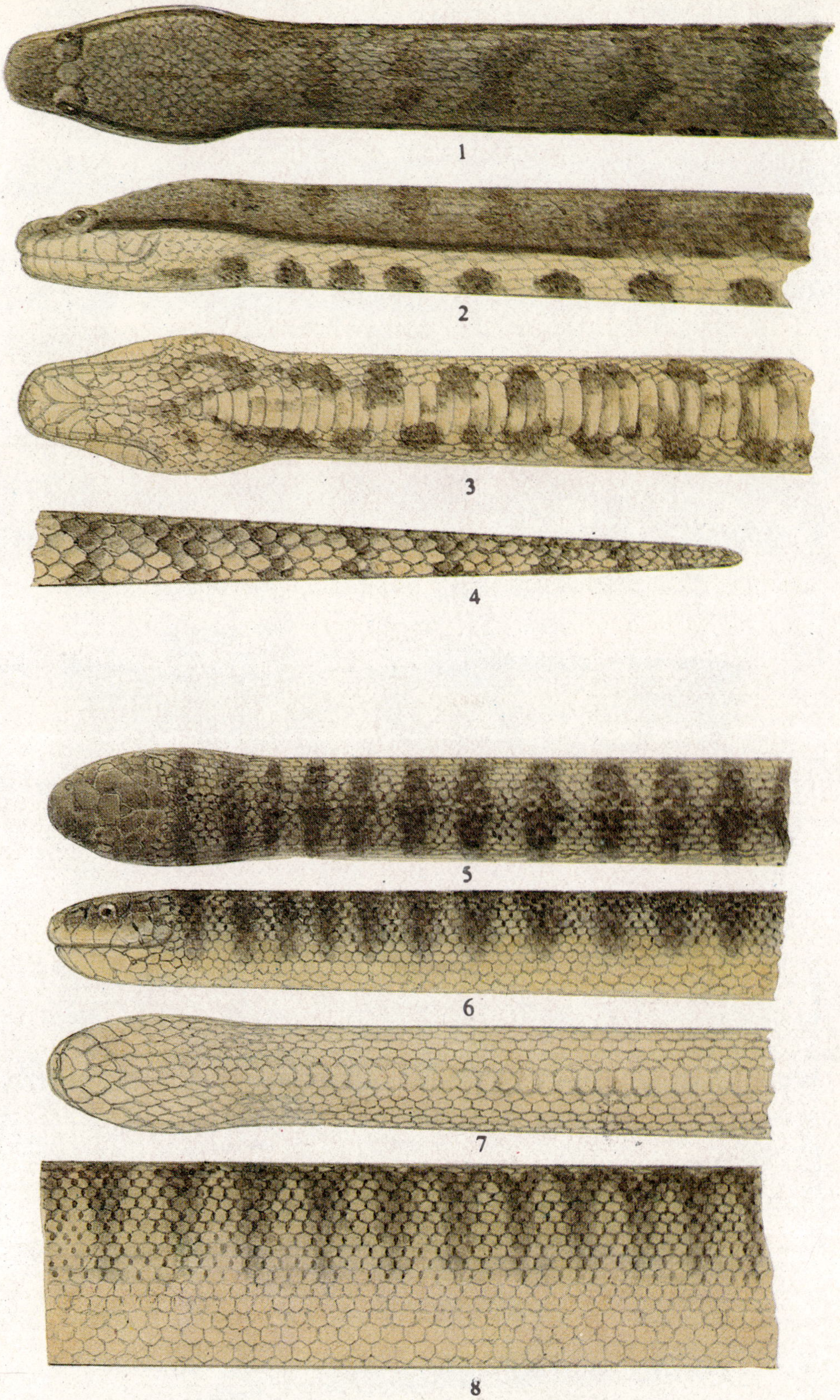

1-4. Dog-faced Water Snake (*Cerberus rhynchops*);
5-8. Shaw's Sea Snake (*Lapemis curtus*) (See pp. 105, 117)

Common Indian Krait (*Bungarus caeruleus*) (See p. 107)

down, the tongue is retracted and the opening of the wind pipe alternately dilates and contracts. With its wide open pink mouth, pouched throat, and the forebody thrown into sigmoid curves ready for striking, the snake presents a formidable appearance.

The habit of striking at the eye, the only object in movement at the level of a tree snake's view of its opponent is noted in the common name for the snake in Tamil which translates as "eye pecking snake".

FOOD. Catholic in diet and is known to take small mammals, birds, lizards and occasionally other snakes and frogs. On seeing the prey the forebody is slowly freed, raised and coiled in a zig-zag manner and at the opportune moment darted forward to unerringly catch the prey just behind the head, drag it off its support and keep it dangling till its struggles cease before swallowing, which may take more than half an hour after capture. Once swallowed the prey is speedily passed on to the stomach by a series of shallow lateral undulations of the body.

BREEDING. The snake is ovo-viviparous and the young are born free from the caul. The appearance of young is usually between March and December. The period of gestation is known to be as long as 172 days or nearly six months (68). Copulation has been observed in June. The number of young varies from 3 to 23 and hatchlings are at birth 260 to 440 mm in length. The young are believed to double their length in the first year and grow at the same rate the second year when they attain a length of 99 to 119 cm. The smallest gravid female measured 118 cm, approximately 2 years of age.

The whipsnake is mildly poisonous, the poison resembling that of the cobra as far as the symptoms are concerned. In man the poison usually has no effect. But occasionally swelling and numbness of the bitten area may result.

SCHNEIDER'S SMOOTH WATER SNAKE

Enhydris enhydris (Schneider)

HARMLESS

LOCAL NAMES. Telugu *Mutta pam, Ally pam.*

SIZE. Longest measured ♂ 890; mm, ♀ 960 mm (Smith).

IDENTIFICATION. The nasals in contact with the rostral, 21 or 23 costals at mid-body and 141-174 ventrals distinguish this snake. A smooth, glossy, moderately thick snake with rounded snout and rather long muzzle. The transverse slit-like nostrils are placed on top of the head. Eyes small, placed high on the face. Iris dotted with golden specks, pupil vertically elliptic.

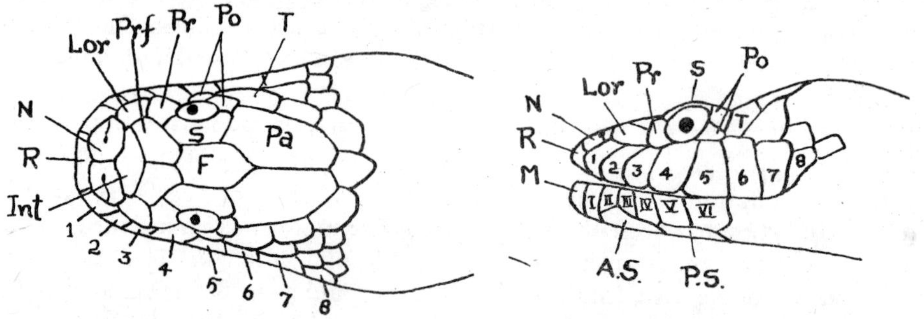

Head Scalation *Enhydris enhydris* (×2)

Key Character: Nasals (N) in contact with rostral (R) shield

COLORATION. Olivaceous-green or brown above, the colour ending abruptly on the 3rd costal row above ventrals and just above the supralabials on the face. A pale stripe on the 8th row above ventral present or absent. Belly lemon yellow. Ventrals demarcated laterally by a dark line. A continuous or interrupted dark line in the middle of the belly may be present or absent.

HABITAT, DISTRIBUTION, and STATUS. An aquatic species frequenting fresh and brackish water. North-east India, from coastal Andhra Pradesh to east U.P., and eastwards to S. China, Indochina, Malaya. Said not to occur south of the Godavary. An uncommon snake.

HABITS. A thoroughly aquatic snake frequenting rivers, estuaries, lakes, marshes and perhaps wet fields. A timid inoffensive snake normally not known to bite when handled. A bite on the hand from a specimen became immediately swollen and the hand started throbbing 15 minutes after the bite and continued to throb for about an hour. There was no bleeding (20).

FOOD. Feeds mainly on fish.

BREEDING. Mating specimens have been seen in October and gravid females obtained in March. Viviparous, 6-18 young are produced.

DOG-FACED WATER SNAKE

Cerberus rhynchops (Schneider)

HARMLESS

LOCAL NAMES. None recorded.

SIZE. Longest measured 1270 mm. ♂ 885; ♀ 1180 mm Smith).

IDENTIFICATION. Frontals partially and parietals entirely broken up. Three or more rows of costals seen in ventral view. Distinguished from other water snakes by the presence of two internasals, 9 to 10 supralabials and 23 to 27 costals at mid body. A rather stout, rough-skinned snake with pear-shaped head, broad at the occiput. Snout narrow in profile with prominent lower jaw giving the snake its forbidding appearance, eyes small, pupil vertical. Tail short rather compressed at base, rapidly tapering to a point.

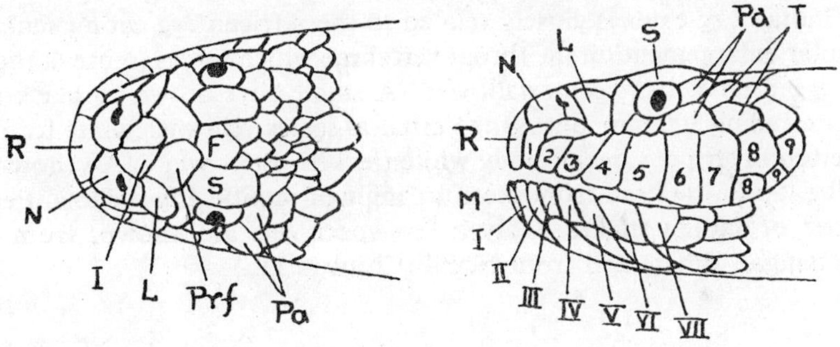

Head Scalation *Cerberus rhynchops* (× 1½)

Key Characters: Frontals (F) partially, Parietals (Pa) entirely broken up. Distinguished from other water snakes by the presence 2 internasals (I) and 9-10 supralabials

COLORATION. Grey above with numerous ill-defined darker crossbars. From midflanks downwards buff. Belly coarsely spotted or dappled with greenish black. Upper lip and chin buff. A conspicuous postocular streak running to the sides of the forebody.

HABITAT, DISTRIBUTION, and STATUS. An estuarine species. Coasts of India and tidal rivers from Sind to Chittagong and eastwards. Common.

HABITS. A fairly common inhabitant of brackish waters of tidal rivers, creeks and estuaries and also the sea littoral. Swims powerfully but is often seen anchored by its tail and swaying in the flow awaiting for fish passing by. In shallow water flicks its tail to frighten fish towards its head. Often climbs on to the low branches of mangroves slipping into the water when disturbed. Inspite of its forbidding appearance it is an inoffensive reptile, biting only under grave provocation. Emits an unpleasant odour under great excitement. It has a curious "sidewinder" movement on land.

FOOD. Feeds exclusively on fish.

BREEDING. Viviparous, gravid females have been obtained in March to August and young obtained in May, July, August. The brood numbers 8 to 26 and measures 177 to 203 mm in length when born. The young of one brood hatched in August were olive above with black crossbars. Black streak through eye to neck, lips and belly white.

INDIAN EGG-EATING SNAKE
Elachistodon westermanni Reinhardt

HARMLESS

The Indian egg-eater is closely related to the African egg eating snake and has similar enlargement on the throat vertebrae which serves to break the shell of the egg after it has been swallowed. A small olive-brown to black snake hardly exceeding 800 mm in length. Vertebral scales yellowish white forming a light vertebral stripe; sides of body white flecked. Belly white. A yellow stripe along the top of the head from snout to angle of mouth. Lips yellow. Perhaps the rarest of Indian snakes. Only a few specimens are known, from India (North Bengal, Bihar) and from Nepal (Chitwan Dt.).

1-3. Yellow-banded Wolf Snake (*Lycodon fasciatus*);
4-6. Banded Krait (*Bungarus fasciatus*) (See p. 109)

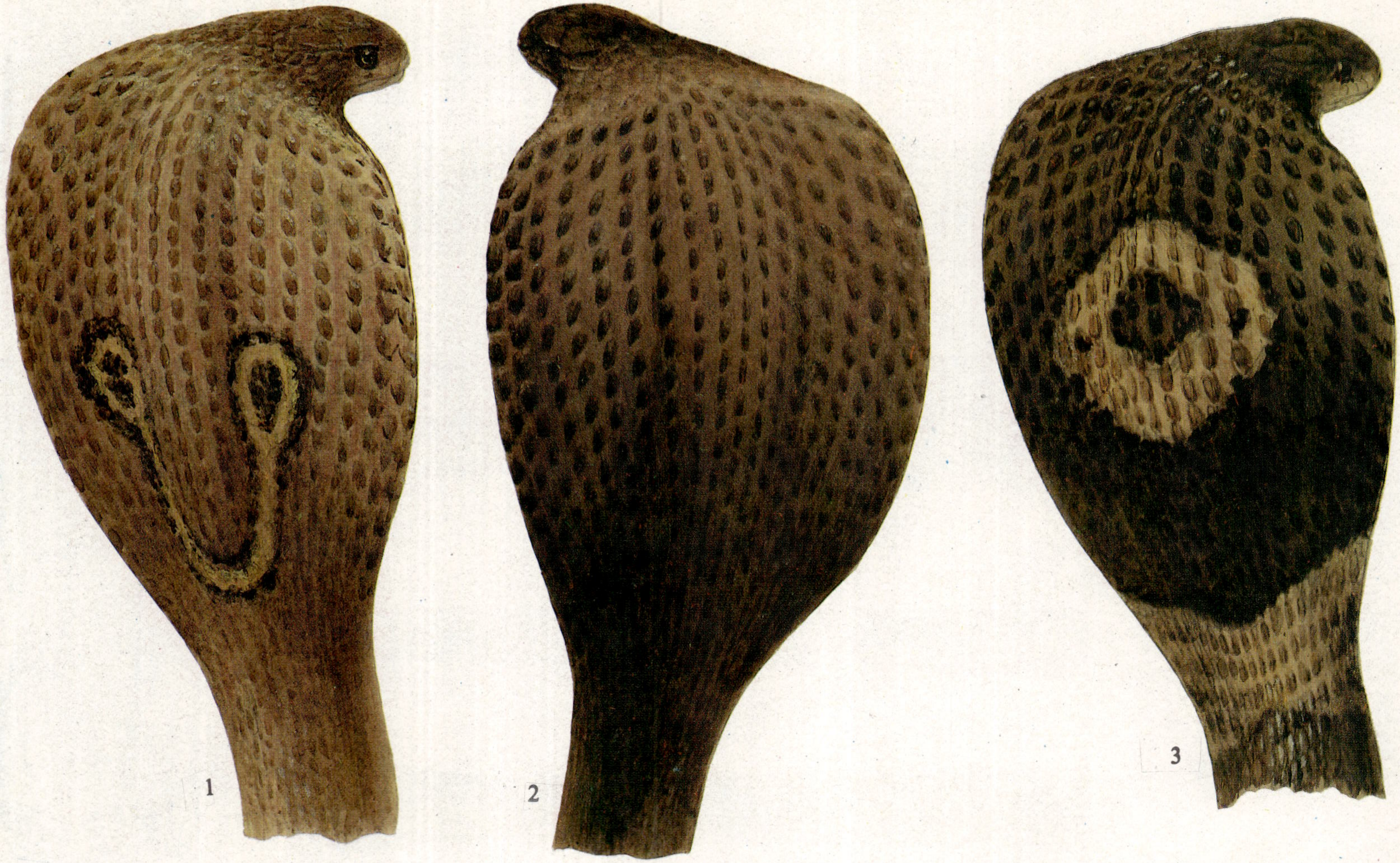

Indian Cobra (*Naja naja*)
1. Binocellate; 2. Black; 3. Monocellate (See p. 112)

COMMON INDIAN KRAIT

Bungarus caeruleus (Schneider)

HARMFUL

LOCAL NAMES. Urdu *Kala gandait*; Gujarati *Kala taro*; Marathi *Manyar, Kandar*; Oriya *Chitti*; Tamil *Kattu viriyan, Yennai viriyan, Panai viriyan, Yettadi viriyan*; Malayalam *Valla pamboo* (Malabar), *Yettadi virien* (Travancore).

SIZE. Largest measured 1.73 m. Specimens over 1.2 m are not common.

IDENTIFICATION. The enlarged hexagonal vertebral scales, the entire subcaudals, the uniformly white belly and the narrow white crossbars on the back, more or less distinctly in pairs, distinguish the species. Body rather long and cylindrical. Neck not evident. Eyes rather small. Scales shiny. Iris black, pupil indistinguishable. Tail short, one sixth to one-eleventh total length.

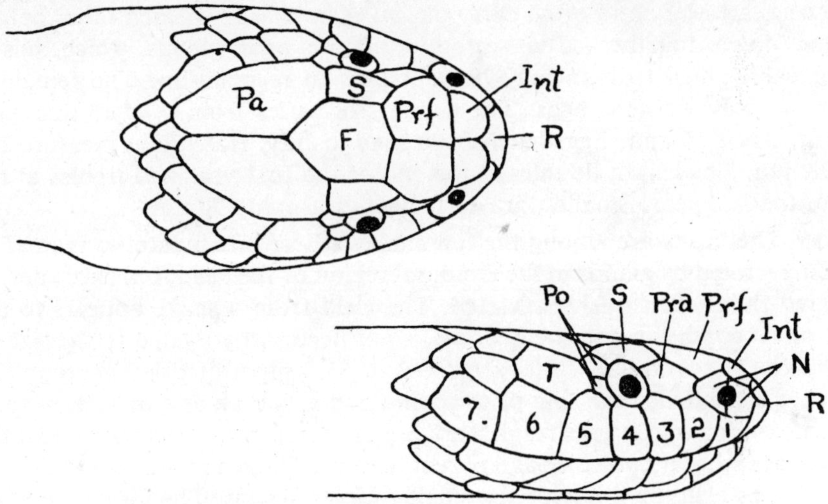

Head Scalation *Bungarus caeruleus* (×2)

For key abbriviations, see Glossary

COLORATION. Lustrous black or bluish black above with paired narrow white crossbars indistinct or absent anteriorly. In young the crossbars are well defined and conspicuous, even anteriorly. In old snakes the white may be in the form of a series of connected spots with a particularly large spot being on the vertebral region. A preocular spot may be present; upper lip and ventrally, white. Two colour varieties are distinguished; with the transverse bars narrow, not usually widening on the sides of the body and without vertebral spots. In the race *sindianus* the bars are distinct, wider on the sides and a vertebral spot is always present.

HABITAT, DISTRIBUTION, and STATUS. Inhabits a wide range of biotopes in peninsular India from Sind (Pakistan) to West Bengal plains and south to the Cape. Sri Lanka. Common.

HABITS. The Common Krait inhabits fields, low scrub jungles and is common in the vicinity of human habitation, often taking up residence inside houses. Frequently found near or in water. It is nocturnal and of a placid temperament biting usually only under provocation. Many instances are on record of people sleeping on the ground being bitten when unknowingly rolling on or placing a leg or hand in their sleep on a Krait moving near by.

FOOD. Feeds mainly on snakes including other kraits. Young feed more or less exclusively on blind snakes of the genus *Typhlina*. Occasionally frogs, lizards, and small mammals are also taken.

BREEDING. The male is longer in total length and has a longer tail. Mating is apparently in February and March. Kraits perhaps from their cannibalistic tendencies are wary and often carry out an examination of each other before settling down together. The secretion of the anal glands which has a disagreeable smell to man may perhaps help in recognition. The female is known to stay with the eggs. The clutch size varies from 6 to 15 and eggs measure 35 × 19 mm. Eggs hatch from May to July. Hatchlings measure 266 to 298 mm. The length doubles at the end of the first year and trebles at the end of the 2nd year. Smallest gravid female measured 890 mm.

POISON. The kraits are among the few snakes whose bite is fatal to man. The poison, secreted by glands in the temporal region of the head, is a clear, amber coloured fluid when freshly secreted. The yield from a snake appears to depend on its condition and the quantity is not necessarily related to the size of the snake. The secretion varies from 0.2 to 51.4 mgm of dried venom. The venom is more toxic than that of the cobra and acts both as a neurotoxin and haemotoxin, paralysing the respiratory centre, and centres concerned with the lips, tongue, throat and voice and phrenic nerves. The red blood corpuscles are destroyed as also the lining of the smaller blood vessels. The major cause of death is asphyxia through paralysis of the respiratory centre. Krait venom is considered to be 15 times more virulent than the Cobra's and the Krait is one of the deadliest among the poisonous snakes of the world (33). The lethal dose for man is considered to be the secretion equivalent to 1 mgm of dried poison.

Symptoms are a 'fiery' pain at the site of the bite which disappears after some time, later violent abdominal pain probably due to haemorrhage and paralysis sets in. The eyelids, and lower lip droop and the person is unable to walk and to breathe. Often there is no immediate reaction and the bite is ignored with fatal results.

Polyvalent Serum should be injected preferably intravenously as soon as possible after the bite. Till medical attention is available the victim should be kept warm and given hot stimulating drinks. Alcohol should not be given.

Death may result in five to twelve hours after the bite.

BANDED KRAIT

Bungarus fasciatus Schneider

HARMFUL

LOCAL NAMES. Bengali *Sankni, Sankhomuti sap*; Oriya *Raona*; Telugu *Bungarum pambah.*

The scientific name of the genus is derived from the Telugu Bungarum = Gold an allusion to the yellow rings on the body.

SIZE. Longest measured 2125 mm but specimens over 1828 mm are exceptional.

IDENTIFICATION. The alternating yellow and black bands are distinctive and the snake can be confused only with the harmless Yellow-banded Wolfsnake, *Lycodon fasciatus (not uncommon in the hills of NE India).* It can be distinguished by the enlarged vertebrals, entire subcaudals, ridged spinal area and blunt tail.

A large snake which attracts attention by its very distinctive and highly ornamental coloration. Head broad and depressed; snout short, eye black, pupil very faintly outlined in yellow. Neck distinct. Body smooth and glossy with a conspicuous ridge down the spine and tail short and ending in a finger-like tip.

COLORATION. Alternately banded with broad canary-yellow and black bands, completely encircling the body. Nape has a large, elongate, black patch, rounded behind. Top of head with a yellow V with its arms passing backwards over the temples to the throat. Lips, lore, chin and throat yellow.

HABITAT, DISTRIBUTION, and STATUS. Prefers open tracts and scrub jungle but often seen in populated areas.

North-east peninsular India, the southernmost record being Hyderabad. Not uncommon in Assam and Bangladesh becomes progressively uncommon westwards. Occurs fairly commonly throughout the Indochinese subregion

HABITS. Largely nocturnal. May lie up in grass, pits or drains during the day. Frequents moist places and the vicinity of water. Extremely sluggish and remains lethargic even under provocation. Most commonly seen during the rains.

FOOD. Feeds mainly on snakes and among those taken are rat snake or dhaman, Indochinese ratsnake *Ptyas korros,* cat snake, chequered keelback, buffstriped keelback. Also skinks, eggs of snakes and occasionally fish. When the prey is longer in length a portion may keep protruding out for a day or two till the portion in the stomach is digested. The prey is swallowed head first usually after it is dead from the effect of the injected poison.

BREEDING. Little is known of its breeding habits. In Burma a female was dug out while incubating a clutch of eight eggs four of which hatched in May (28). Young measure 298 mm to 311 mm at hatching. Believed to become adult in the 3rd year of life when about 914 mm length.

The snake as other kraits is poisonous but there is no record of a bite by this species. The poison is said to be less virulent than Cobra poison though a bullock struck by it is reported to have died in about 20 minutes after being bitten.

MACCLELLAND'S CORAL SNAKE

Callophis macclellandi (Reinhardt)

HARMFUL

LOCAL NAMES. None recorded.

SIZE. Longest recorded 2 ft 8 in. (♂ 705; ♀ 840 mm. Smith).

IDENTIFICATION. The broad enamel-white band across head distinctive. The 13 rows of costals, suture below nostril passing to 2nd labial; 7 supralabials with 5th and 6th touching temporal shield distinguish the species from other Indian snakes. Head flattened and broad. Body cylindrical, moderately robust and of uniform thickness throughout. Tail short.

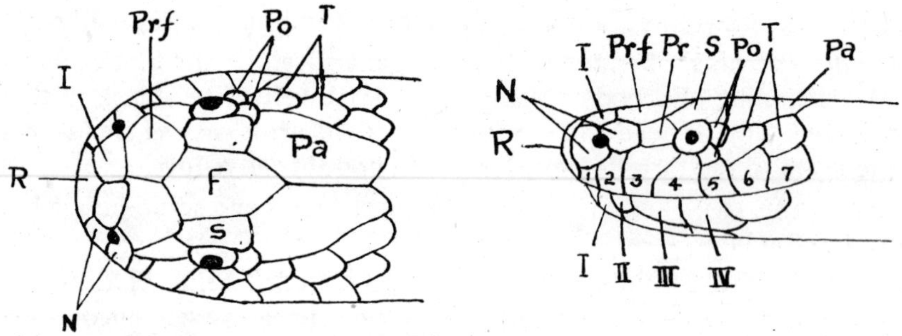

Head Scalation *Callophis macclellandi* (×2½)

Key Characters. Suture below nostril (N) passing to 2nd labial. 7 supralabials, with 5th & 6th touching temporal (T) shield.

For key to abbreviations, see Glossary.

COLORATION. Many colour forms, the most widely distributed being red dorsally varying in richness from bright strawberry to cherry, rarely purplish. Regular narrow, black transverse bars across back which may or may not reach the belly. Head shiny black with a sharply defined broad ivory white, rarely cream-coloured, cross band behind eye. Belly saffron, blotched with black.

A variety with a black vertebral stripe and transverse bands reduced or absent occurs in the eastern Himalayas up to Khatmandu.

HABITAT, DISTRIBUTION, and STATUS. From the western Himalayas to

Banded Krait (*Bungarus fasciatus*)

Photos: M. Krishnan

Common Indian Krait (*Bungarus caeruleus*) (See pp. 109, 107)

Indian Cobra (*Naja naja*)
Albino

Photos: M. Krishnan

Binocellate (See p. 112)

southern China and Taiwan. The typical form is common in the hills of Assam and eastwards. The variety *univirgatus* is abundant in Sikkim. Eminently a jungle as well as a hill species.

HABITS. Little known. Prefers heavy forests between 1200 and 1800 m but may occur at lower elevations. An inoffensive and sluggish species said to feed mainly on snakes.

BREEDING. Apparently breeds during the rains, gravid females obtained in July (Burma), August (Assam) had 14 and 6 eggs respectively, the latter with embryos 2 to 3 cm in length.

No information is available on the virulence of its poison.

A fairly widespread but uncommon species in India is the SLENDER CORAL SNAKE (*Callophis melanurus*) distinguished from other Indian Coral snakes by the presence of only six supralabials. Light brown above with the centre of each scale speckled with brown which form longitudinal lines down the whole length of the body. Head and neck black above spotted yellow, a pair of spots on occiput distinct. Tail with 2 black rings one at base, the other near the tip. Below red. The snake when disturbed curls its tail and exhibits the red under surface. Occurs in peninsular India and Ceylon.

Indian Cobra

Naja naja (Linn.)

HARMFUL

LOCAL NAMES. In most parts of India derivatives of the Sanskrit *Nag*; Bengali *Naga gokurra* (binocellate form), *Keauthia* (monocellate form); Pushtu *Cha-jitiwalla*; Tamil *Nalla pambu, Naga pambu*; Kannada *Nagara havu*; Malayalam *Moorkan, Surpam*; Singhalese *Naya*.

SIZE. Longest measured 2250 mm. Usually from 1371 to 1625 mm.

IDENTIFICATION. The cobra can be immediately distinguished from other land snakes by the presence of a small 'cuneate' scale between the 4th and 5th infralabials. Rarely two may be present and very rarely the cuneate may be absent. Another distinguishing character is the preocular touching the internasal, a character seen in two other species of Indian snakes also but the cobra can be separated from these in having the 3rd supralabial in contact with the eye.

The hood is formed by the elongated ribs of the 3rd and the following 27 vertebrae, the 9th on the left and 10th on the right are the longest, the preceding and succeeding ribs shorten progressively giving an oval outline to the expanded hood. At rest the ribs lie along the length of the body, the overlying skin is but loosely attached. When erect the dorsal skin is stretched making the hood markings conspicuous and the head bent strongly at the atlas (1st vertebra) is carried at right angles to the hood. The hood when dilated is diagnostic, more so when the markings are visible. The markings may be absent and in death the hood may not be demonstrable. The King Cobra has a well developed hood and many other snakes have the ability to flatten the neck area to a more limited degree.

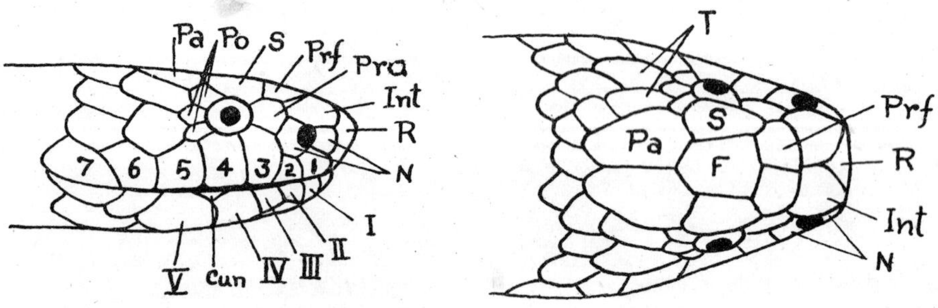

Head scalation *Naja naja* (nat. size)

Key Characters: Cuneate scale (Cun) between 4th and 5th infralabials (lower lip) shields, Preocular (Pro) touches Internasal (Int).

3rd Supralabial (upper lip shield) touches eye.

For key to abbreviations, see Glossary.

Head depressed with short, rounded snout. Nostrils large, pupil round, an obvious swelling at the temporal region over the underlying poison glands. Head shields glossy, body with a more or less distinct groove down the spine.

COLORATION. Extremely variable in coloration and markings. Three races are recognised on the basis of the hood pattern: The 'spectacled' or binocellate Cobra of peninsular India (*Naja naja naja*) yellowish, brownish or black above with or without a black and white mark on hood, a black and white spot on the inside of the hood with one or two black crossbars below hood. Sri Lankan and south Indian cobras are usually of shades of brown with well-defined hood marks. Cobras from the north are more often black and the hood pattern may not be well defined or may be absent.

Monocellate Cobra (*Naja naja kaouthia*) differs in having only a single yellow or orange O-shaped mark on the hood. General colour olive, brown or black. This is the common Cobra of eastern India and eastwards of India.

The Black Cobra (*Naja naja oxiana*) occurs in the extreme northwest. Light grey or brown above when young with dark crossbars. Adult brown or black. Uniform.

HABITAT, DISTRIBUTION, and STATUS. Eclectic in habitat. Absent in arid deserts and in the hills above 1800 m. Occurs from Transcaspia in the north, through Indian subcontinent to southern China in the east and to the Philippines in the south. Andamans. Sri Lanka. Common.

HABITS. Found almost anywhere, in heavy jungle, open cultivated land, in populated areas where old masonry constructions form ideal refuge. White ant nests, holes in the ground or the tangle of roots at the base of a tree are particularly favoured. Frequently found near or in water and is a strong swimmer.

Usually not aggressive and often exceedingly timid but occasionally fierce and aggressive when disturbed. Young are much more dangerous than adults being more easily excited and ready to strike repeatedly and with determination. When alarmed it adopts the well known pose with erect forebody and spread hood. The height to which the forebody is raised is approximately one-third the total length of the snake and forms the effective striking range. Whilst thus poised the snake sways backwards and forwards hissing in an explosive manner brief and high pitched during inhalation and longer, louder, lower pitched and intermittently explosive during exhalation. The throat is pouched, more so, during exhalation and the whole body is inflated. The tongue flickers in and out during inhalation and exhalation. The bite is often a mere snap but sometimes bites and hangs on and the jaws have to be forced open. Occasionally when the snake misses, the poison is ejected as a spray by the forceful thrust of the lunging snake. Usually more active and alert at night though hunts for food during the late afternoon and early evening.

FOOD. Feeds principally on rats, frogs and toads. Also takes birds, lizards, other snakes including other cobras and is an inverterate egg stealer. Eggs are swallowed whole and digested in about 48 hours (10).

BREEDING. Mating has been seen in January and the majority of eggs are laid in April/May but clutches have been obtained up to August. The period of gestation is about sixty two days but may extend considerably. Eggs hatch in 48 to 69 days. Twelve to twenty two, in one instance 45 (36 fertile) eggs are deposited at a time. The eggs are soft-shelled elongate oval measuring 49 × 28 mm. The parents cohabit before pairing and the eggs are guarded by one or both. Both parents known to incubate. Hatchlings measure 250-280 mm at birth. The poison glands are active from birth.

POISON APPARATUS and POISON. Usually two fully operative canaliculate fangs on each side. These are shed singly at intervals. Fangs about 7 mm in length are small compared to viperine fangs but are more solid. The bore of the fang opens widely at the base and by a small aperture at the tip. The poison glands are analogous to the parotid salivary glands in mammals and have the shape and size of an almond kernel.

The venom is a clear, viscid fluid resembling olive oil in appearance and consistency which solidifies into an amorphous mass. The amount secreted varies with age, vitality and temper of the animal and the average discharge at a bite is about 211 mgm in dry weight. Comparative data on the basis of experiments on other animals gives the lethal dose for man as 15 to 17.5 mgm for a weight of *c.* 60 kg. The poison can be, however, swallowed without ill effects provided there are no internal ulcers. The poison acts mainly as a neurotoxin and blood and cell destroyer. The neurotoxin paralyses the respiratory centre and is the chief cause of death. Other effects are loss of clotting power of the blood and destruction of red blood cells.

The symptoms produced in man, start with a stinging or burning pain accompanied by swelling and oozing of blood-stained serum. The constitutional effects are a gradual but rapidly advancing paralysis commencing with the legs, the neck droops, the muscles of the tongue, lips, and throat, are affected and speech becomes difficult. The lower lip falls and allows saliva to dribble, swallowing becomes difficult or impossible. Breathing becomes difficult, laborious and stops. Other symptoms are vomitting and haemorrhage from the various orifices of the body.

It does not necesarily mean that the bite of a cobra is fatal at all times, depending as it does on the quantity of venom injected, the natural resistance of the victim, the condition of the snake and various other factors. Records indicate that case of recovery from a bite is equal to if not more than cases of death and there is always hope however serious the symptoms.

The Haffkine Institute's polyvalent serum is fully effective even when symptoms are far advanced.

King Cobra or Hamadryad

Ophiophagus hannah (Cantor)

HARMFUL

LOCAL NAMES. Oriya *Ahi raj*; Bengali *Sankha chur, Sha-khamuti*; Assamese *Fetty sap*; Shan *Gni son-an*; Karen *Gni thaw*; Tamil *Krishna nagam, Karunagam;* Malayalam *Krishna sarpam, Karinchathii*; Kannada *Kalinga havu, Nagin, Kalinagin*.

SIZE. A 18'4" (*c.* 5.5 m) specimen has been recorded. A specimen obtained in Singapore Island had a length of 4.77 m (15'7") and weight of 12 kg (26½ lb.) (14).

IDENTIFICATION. The presence of a pair of occipital shields and costals in 19-19:15:15 rows are distinctive. The third largest snake in the Indian region. The body is fairly robust and the scales glossy. The 'hood' is relatively less dilatable than in the cobra. Head flat; snout rounded and eyes moderately round with round pupil.

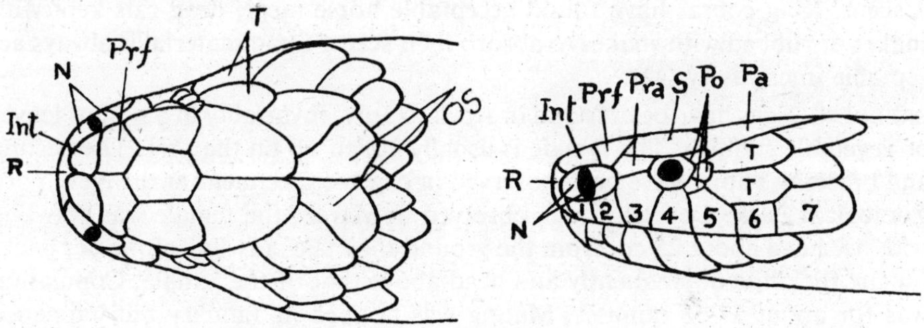

Head Scalation *Ophiophagus hannah*

Key Character: OS Occipital Shields

COLORATION. Adults blackish brown to light brown with 32 to 43 lighter bands round body and 11 to 13 round tail. Bands conspicuous in juveniles and gradually become obscure with age. Bands become more evident under excitement. Head olivaceous-brown. Throat creamy to dull orange merging into a dark mottling which becomes, further back, a uniform slate or brown.

Hatchlings intense black with pure white bands. Head black with three white bars, the last with an oblique stripe passing to the side of the throat.

HABITAT, DISTRIBUTION, and STATUS. In the Peninsula it occurs in the dense forests of the Western Ghats and in the forests of the hills, plains and estuaries of Orissa, Bengal and Assam. There is a single record of its occurrence at Lahore, Pakistan and another at Deesa, Palanpur, far from its usual haunt. Peninsular India to the Himalayas, Assam and eastwards to the Philippines. Andaman Islands, Hongkong. It is not a common snake in India.

HABITS. Inhabits dense jungles of hills or their near vicinity in peninsular India. The Hamadryad has earned for itself an unenviable reputation for aggressiveness and courage and is probably unique among Indian snakes for these traits. There is no doubt that the snake will sometimes attack without provocation. Several instances are known of jungle paths being closed to traffic owing to a brooding female or her mate (?) attacking passers by. Aggressiveness or timidity is dependent on individual temperament and when molested one may attack with great determination and another may speedily slink off.

Largely diurnal, its movements are singularly rapid but not rapid enough to overtake a running man as has been alleged (32). Like the cobra, the hamadryad erects its forebody for about one-third its length and spreads the hood. The bite is a determined one and the snake holds on tenaciously, pumping in the poison with a chewing motion of its jaws.

FOOD. The staple diet is snakes. All species, including other poisonous snakes being taken. Occasionally monitor lizards are eaten. The primary feeding cue is scent. King cobras have found acceptable horse meat, dead rats kept with snakes or rubbed with snakes to absorb their scent. Dead material is always acceptable in captivity (42).

BREEDING. Eggs have been found in April to July, invariably in a nest of leaves or vegetable rubbish. The female is usually coiled up on the nest. The nesting and breeding habits have been observed in captive specimens at the New York Zoological Park. Courtship was observed in March, the female was crawling with her head about 25 cm from the ground the male crawling along her back, tongue flicking out frequently and head above that of the female. Copulation was for about 57-58 minutes. Mating was noticed in January but happened mainly in March. In April the female started a nest of leaves hooking in the leaves with a loop of her body and piling them up. Sand was also brought into the nest heap by loosening it with tilted head and raking it in with a loop of the body. A chamber was made inside the pile by the snake revolving in the middle of the structure. Additional material was added on. In one instance a bunch of leaves was literally carried in a loop of the body, the holding coil being held about two inches above the ground. The nest was completed in three days. The eggs are laid in the central chamber and covered almost completely with a layer of litter and the female coils on top of the leaf-covered clutch. The eggs, which measure 59.4 × 34.44 mm and weigh 40.85 gm on an average, are laid five to six weeks after mating and up to 51 eggs have formed a clutch.

During incubation eggs increased on an average, 66.1% in weight, 8.1% in length, and 26.2% in width. The first egg hatched 10 weeks and a day after laying and the last egg to hatch was eleven weeks and 3 days after laying. Hatchlings measure *c.* 50 to 52 cm at birth.

POISON. The fresh poison is a clear, tasteless, slightly acidic fluid. Laboratory experiments indicate that quantity equivalent to ten lethal doses to man could be discharged at a bite. The virulence is less than that of the Cobra. Considering the size of the snake and the vehemence of its attack, a sublethal dose is unlikely from a bite. Death has been recorded in 15 to 20 minutes. No antivenin is manufactured in India but is available in Thailand.

It has lived in captivity for over 12 years.

SHAW'S SEA SNAKE

Lapemis curtus (Shaw)

HARMFUL

LOCAL NAMES. None recorded specifically apart from the general name for sea snakes as *Samudra sáp* (*nág*), Hindi; *Kadal pambu,* Tamil and Malayalam.

SIZE Maximum recorded 846 mm (*c.* 2 ft 10 in.).

IDENTIFICATION. The only Indian sea snake with parietal shields on head broken up into smaller shields. Ventral shields slightly enlarged anteriorly, 130 to 219 in number. A small snake with stout cylindrical body, compressed posteriorly.

COLORATION. Olive-green turning to pale yellow above, with ill-defined, dark greenish brown or black crossbars 45 to 55 in number, the first one on the nape. Young with bars in the form of rings.

DISTRIBUTION and STATUS. Persian Gulf to Malay Archipelago. Common along the Malabar and Coromandel (Tamil Nadu) coasts.

HABITS. A graceful sea snake. In appearance more like an elongate fish. Infrequently surfaces for air.

FOOD. Fish, species not determined.

BREEDING. Viviparous; young one to four at a birth between May and August. Young at birth 330 to 355 mm in length. Females mature at about two years of age.

POISON. Similar in action in experimental animals to that of the cobra, but the respiratory failure is more pronounced. No report of humans being bitten.

VALAKADIYAN or HOOK-NOSED SEA SNAKE

Enhydrina schistosa (Daudin)

HARMFUL

LOCAL NAMES. Bengali *Hoogly patee* (Sunderbans area); Tamil and Malayalam *Valakadiyan* (=net biter).

SIZE. Average length 915 to 1220 mm. Maximum recorded 1580 mm (Smith).

IDENTIFICATION. The oar-shaped tail distinguishes this and other sea snakes. The well marked furrow on the chin and rostral projecting well below the lip separate this species from other sea snakes. A moderately large snake, with elongate, robust, subcylindrical forebody. Compressed posteriorly. Head large, snout short, bowed in profile. Eye with dull green iris. Neck hardly apparent.

COLORATION. Variable. Young bluish or bluish grey with well marked black rings often broadened vertebrally. With age bands disappear or may remain as bands dorsally. Old adults bluish or bluish gray above, yellowish below.

HABITAT, DISTRIBUTION, and STATUS. Coastal waters from Persian Gulf to New Guinea. Abundant on both coasts of the Indian peninsula.

HABITS. The commonest Indian sea snake being very numerous all along the coast and ascending considerable distances on the tidal rivers. It has been obtained at Tolly's Nullah, Calcutta, 80 miles from the sea. Though virulently poisonous, they are conspicuously gentle creatures and inspite of their abundance and being handled by fishermen while removing them from nets, instances of bites from this species in India are rare but has been reported from coastal area of Tamil Nadu. Cases of bites by this snake have been more commonly reported from Malaya. In one instance death occurred in 20 hrs. The poison is said to be ten times more potent than cobra venom.

FOOD. They live entirely on fish.

BREEDING. Viviparous, four to nine young are born, the season being from February to May. Young at birth about 279 mm in length. The young double their length in a year and are believed to attain maturity at 3 years. The smallest gravid female obtained measured 965 mm.

YELLOW SEA SNAKE

Hydrophis spiralis (Shaw)

HARMFUL

LOCAL NAMES. None recorded specifically.

SIZE. Most adults range between 1370 and 1675 mm. Specimen over 1830 mm are unusual. A specimen 2745 mm has been reported.

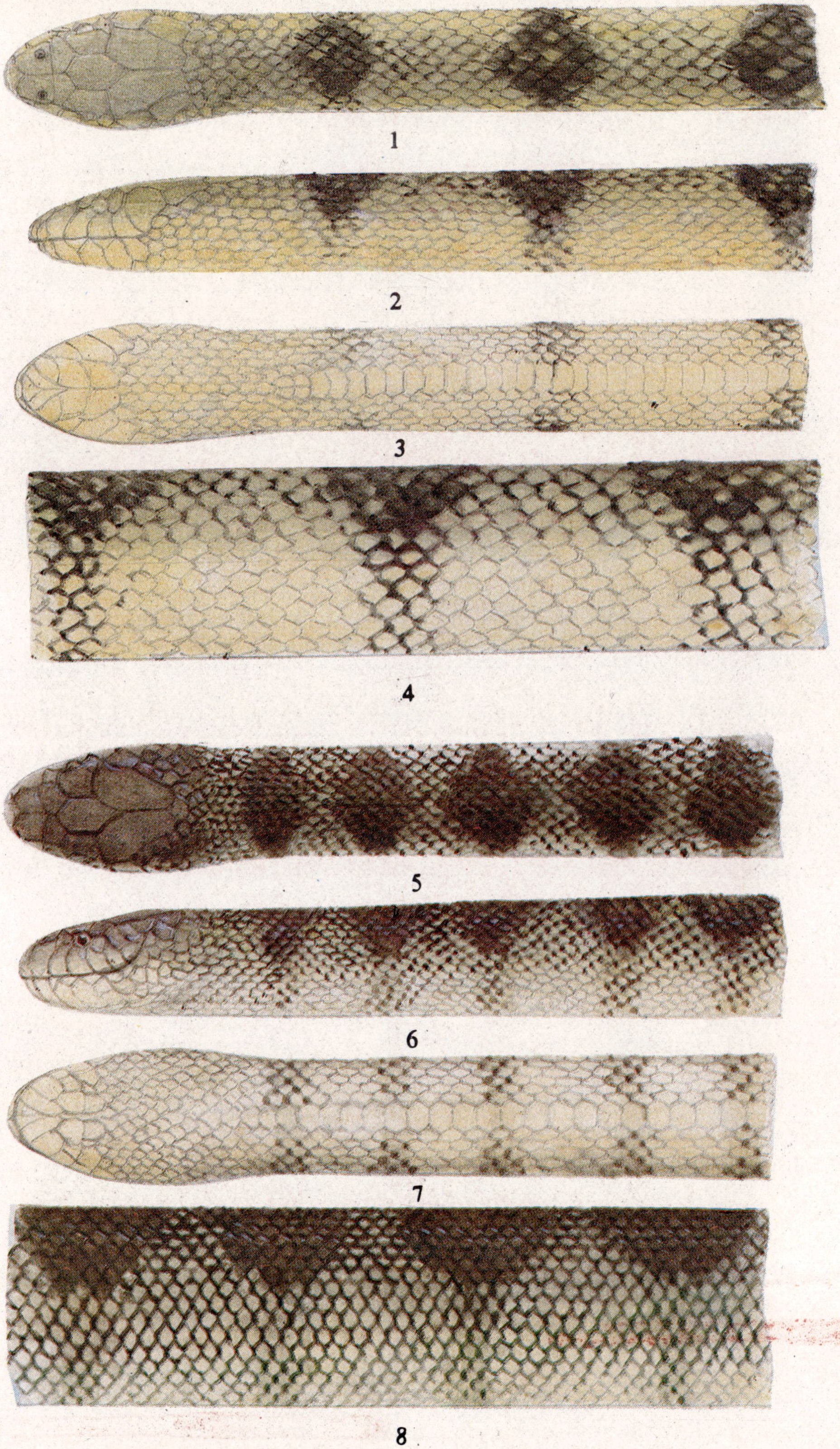

1-4. Yellow Sea Snake (*Hydrophis spiralis*); 5-8. Chittul (*Hydrophis cyanocinctus*) (See pp. 118, 119)

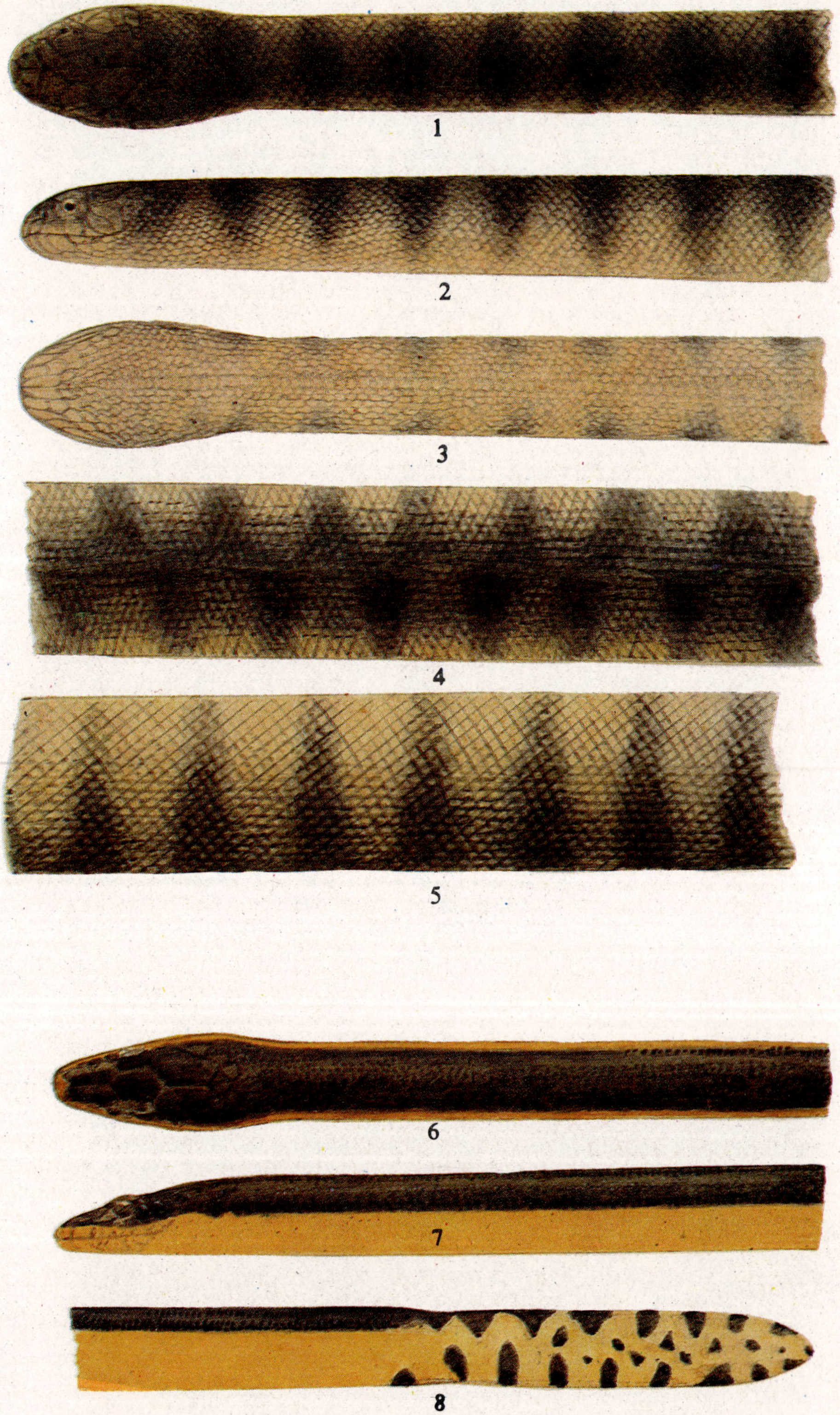

1-5. Valakadiyan or Hook-nosed Sea Snake (*Enhydrina schistosa*); 6-8. Flat-tail (*Pelamis platurus*) (See p. 118, 120)

IDENTIFICATION. The narrow bands taken with the imbricate costals in 25 to 31 rows two head-length behind the head and 29 to 36 at the greatest girth distinguish this from other sea snakes. The largest among the sea snakes growing exceptionally to over 2 ½ metres. Head moderately large and broad. Eye small, neck hardly apparent. Body elongate, cylindrical, compressed in the posterior three-fifths.

COLORATION. Head in young black with more or less distinct yellow horseshoe mark on the crown. The black fades with age and the head may become light olivaceous above, yellow ventrally with 34 to 70 variable shaped black bands or bars.

HABITAT, DISTRIBUTION, and STATUS. Persian Gulf to Malay Archipelago. Abundant in the Persian Gulf and along the coast to Karachi, uncommon on the west coast of the Indian peninsula and common on the east coast.

HABITS. A strong and active swimmer going far up tidal rivers.

FOOD. Feeds largely on eel-like fishes.

BREEDING. Viviparous, gravid females have been obtained in January and June. Five to fourteen in one brood. Young are about 408 mm at birth.

CHITTUL

Hydrophis cyanocinctus (Daudin)

HARMFUL

LOCAL NAME. Bengali *Chittul.*

SIZE. Average adults range between 1220 and 1525 mm. Longest examined 1885 mm (Smith).

IDENTIFICATION. Costals 27-36 at neck and 38-49 at greatest girth. Ventrals 296 to 398. 3rd, 4th and 5th labials touch eye but this character is not common. A moderately large species with broad slightly depressed head. Snout long and projecting. Body cylindrical anteriorly, compressed in the posterior three-fifths.

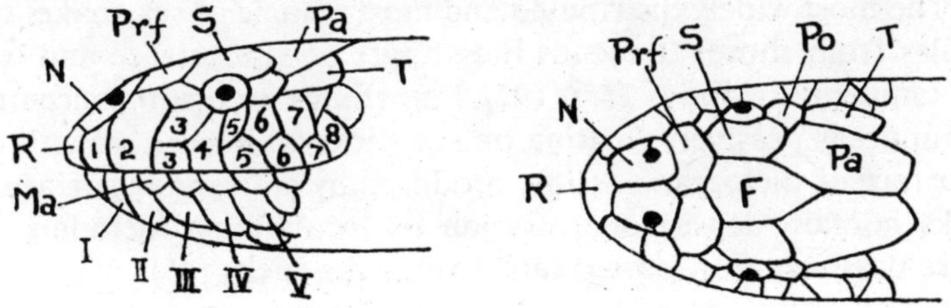

Head Scalation *Hydrophis cyanocinctus* (nat. size)

Key Character. 3, 4, 5 labials touch eye

COLORATION. Head in young as in *spiralis.* Turns olivaceous with age. Body olivaceous dorsally, yellow ventrally with 41 to 70 black bands varying in pattern. 3 young born in the Madras Aquarium were saffron yellow with black stripes. The male more markedly banded with burnt sienna than the female, on the lower jaw, has thicker tail, and strongly keeled scales.

HABITAT, DISTRIBUTION, and STATUS. Persian Gulf to Japan and Papuasia. The commonest sea snake from Persian Gulf to Baluchistan. One of the commonest sea snakes along the coasts of the Indian peninsula.

FOOD. There is no information on the food but snakes kept in the Madras Aquariam were fed on chopped fish.

BREEDING. Viviparous. A specimen in captivity gave birth to three young in December. The young started feeding six days after birth. A female killed in May had 30 developing eggs.

FLAT - TAIL

Pelamis platurus (Linn.)

HARMFUL

SIZE. Longest measured 880 mm (Smith).

IDENTIFICATION. Costals broader than long in 40 to 54 rows on neck. The coloration is also distinctive. A small snake with elongate, depressed head. Snout long, neck fairly evident. Body compressed with a sharp ridge along the back deepest near midbody. Belly forming an obtuse keel.

COLORATION. Very variable. Three colour forms on the Indian coasts. Head black above, the upper lip yellow. Body may be black above brown below with a yellow stripe in between or a black vertebral stripe, wavy or broken into spots, yellow on the sides and below. The colour of the snake may be apsomatic.

HABITAT, DISTRIBUTION, and STATUS. East Africa to Pacific coasts of America. Moderately common around the Indian coasts.

HABITS. The most widely distributed and most pelagic of sea snakes often seen many miles from shore. It spends long hours on the surface and is active in water of temperature above 25°C (92). Sometimes occurs in concentrations of several hundreds passively floating on sea slicks which are smooth long lines on the surface of the ocean possibly produced by converging surface currents. The slicks contain dense concentration of local fauna including fish. The snakes are passively transported into and by the slicks (45).

FOOD. Feeds on fish, which are swallowed head first after they are dead. It is said to have a fishy odour (46).

BREEDING. Viviparous, gravid females with ready to emerge young have been obtained in March.

RUSSELL'S VIPER

Vipera russelli (Shaw)

HARMFUL

LOCAL NAMES. Hindi *Daboia*; Sindhi *Koraile*; Bengali *Bora, Chandra bora, Uloo bora*; Gujarati *Chitalo, Khadchitalo;* Marathi *Ghonas*; Kannada *Mandalatha havu, Kolakumandala*; Malayalam *Mandali, Ruthamandali*; Tamil *Retha aunali, Kannadi virian*; Telugu *Katuka rekula poda*; Burmese *Mwe lewe*; Singhalese *Tic polonga.*

SIZE. Specimens 1500 mm and over exceptional. Usually about 1200 mm. Maximum size recorded 1675 mm.

IDENTIFICATION. Identified by a combination of characters; head covered with small scales and without shields, 27 to 33 costals at midbody;subcaudals divided. Body massive, cylindrical, narrowing at both ends, head flat, triangular with short snout, large gold flecked eyes with vertical pupil and large open nostrils. Neck constricted. Belly rounded. Tail short about 1/7th total length.

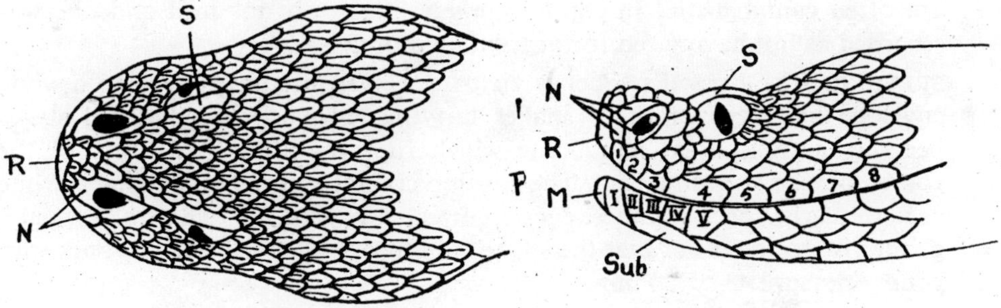

Head Scalation *Vipera russelli* (nat. size)

COLORATION. Ground colour brown of varying shades with three series of large ovate spots, one vertebral and two costal. Spots brown in the centre and margined successively by black and white or buff. The dorsal spots may coal esce and the side spots may be broken up. Smaller spots may occur between the dorsal and side spots. Head with distinct dark patch behind. A dark streak, margined with white, pink, or buff behind eye. A dark stripe from eye to lip. A conspicuous white, buff or pink line from gape converges to form a V above snout. Lips white, whitish or pink. Belly white, whitish or yellowish with a few dark half moon marks on the margins of the anterior ventrals.

HABITAT, DISTRIBUTION, and STATUS. Widely distributed but prefers open country. Indian subcontinent from Baluchistan in the west and Kashmir in the north to the eastern Himalayas and eastwards to Burma, Thailand, Indo-china, Formosa. Indo-Australian Archipelago and Sri Lanka. Usually in the plains but has been recorded up to 2100 m in south India, and 1800 m in the western Himalayas. In some parts of the country it is very common and rare in others. It is abundant in the Punjab, very common along the west coast and its hills, and in south India generally and up to lower Bengal. Uncommon to rare in the Ganges Valley, North Bengal and Assam. Abundant in Burma. It is not uncommon in inhabited areas, the attraction being the rodents commensal with man.

HABITS. Normally sluggish and does not strike readily unless irritated when it bites with great malice. Usually it contends itself with hissing which once heard is not easily forgotten, the volume of sound exceeding that produced by any other snake. When striking it hurtles itself forward and may even leave the ground. The bite may be either a snap or the snake may hold on for many seconds. Largely nocturnal, its movements are slow, never exceeding a crawl and if disturbed often prefers to maintain its ground angrily hissing with heaving sides. The young are more prone to be aggressive and to bite. The main food is murid rodents. In capitivity it has taken, in addition to rats and mice, squirrels, shrews, kittens, small birds, calotes, lizards and frogs. The young are often cannibalistic. In captivity many adults do not feed and one was recorded as not having fed for nearly five months.

BREEDING. The Russell's Viper is viviparous. Fertilised eggs develop a white envelope like eggs of other snakes but this envelope in advanced stages becomes a transparent membrane which ruptures prior to delivery or the young may be born in a caul. The envelope in unfertilised eggs remains white as in the early stages and these eggs are frequently voided along with the young giving rise to the belief that the snake is both ovi- and vivi-parous. Sacs with young measure 43 × 20 mm.

Gravid females have been obtained in all months of the year. Young are born between May and November with a peak period of birth in June and July. The gestation period exceeds six months. One of the most prolific of Indian snakes frequently producing thirty to forty young. The maximum recorded is sixty three (79) but instances of a single foetus or less than twenty are known. Length at birth varies from 215 to 260 mm. Smallest gravid female recorded 1,015 mm about three years in age.

POISON. The fangs attain their maximum size in this the largest of Indian Vipers. Average size is about 16 mm. There are two fangs to a side with 5 or 6 reserve fangs lying behind. Fangs are movable and the fangs can be erected when the mouth opens. The poison glands are small and present a corrugated appearance.

The venom is transparent, acidic in reaction and tastes like gum arabic. When dried, retains its toxicity indefinitely and is readily soluble in water. The total yield may be about 145 mgm and about 72 mgm may be injected at a bite, considerably in excess of the 42 mgm thought to be the fatal dose for man. The

Female on nest
King Cobra or Hamadryad (*Ophiophagus hannah*)

Photos: R. Whitaker

Juvenile (See p. 115)

Saw-scaled Viper (*Echis carinatus*)

Photos: M. Krishnan

Russell's Viper (*Vipera russelli*) (See pp. 124, 121)

poison acts as a depressor of the vasomotor centre and a destroyer of the blood. The blood pressure drops and heart weakens. The red blood corpuscles are destroyed, the clotting power of the blood is reduced and the lining of the blood vessels destroyed leading to extensive internal haemorrhages with pain and vomiting and bleeding from the body's openings. In experimental animals massive doses of the venom result in extensive clotting of the blood and death in a few minutes owing to action of a principle that clots blood and is only active in high concentrations.

The symptoms in man are intense burning or stinging pain at the site of the bite, rapid swelling of the area, and constant oozing of a thin bloody serum from the puncture. The pulse becomes rapid and weak and breathing rapid, irregular, accompanied by muscular weakness, nausea, and vomiting. Pupil becomes dilated and insensitive to light. Unconsciousness may result. The skin becomes cold often bedewed with sweat. Bleeding from body openings and internal haemorrhages occur. Death from cardiac or respiratory failure or speticimia may occur in 1 to 14 days or even later. The Haffekine Institute Polyvalent Scrum is an effective antidote to the poison.

SAW-SCALED VIPER

Echis carinatus (Schneider)

HARMFUL

LOCAL NAMES. Pushtu *Phissi*; Sindhi *Kuppur, Janndi*; Hindi *Afai*; Marathi *Phoorsa*; Kannada *Kallu havu*; Tamil *Viriyan pamboo, Surutai viriyan*; Malayalam *Churuta*; Gujarati *Tarachha, Zeri padkoo udaneyn.*

SIZE. Usually less than 457 mm in length, rarely attains 610 mm. Longest recorded 788 mm.

IDENTIFICATION. Distinguished from other Indian snakes by the absence of shields on the head, the broad ventrals covering the whole belly and the undivided subcaudal shields. Body cylindrical, short and stout, rough from the

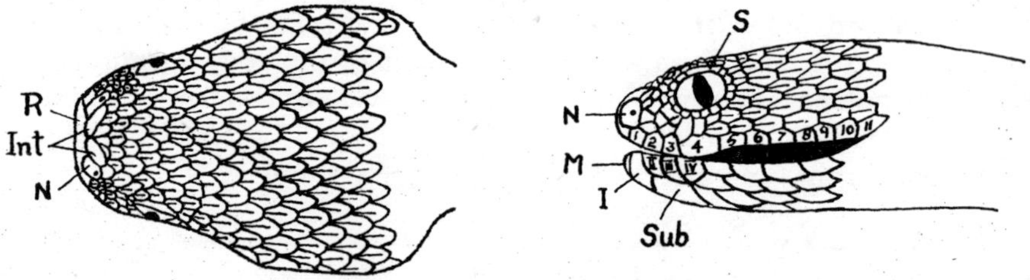

Head Scalation of *Echis carinatus* (×2)

serrated flank scales, tapering towards both neck and vent; neck distinctly constricted. Head subovate with short rounded snout. Eyes large, Iris golden yellow, pupil vertical. Tail short.

COLORATION. Colour and pattern varies considerably. Pale brown, buff or tawny with dark brown or even blackish markings in the form of dark edged spots in a vertebral series connected to a light coloured inverted U- or V-shaped flank mark enclosing a dark area connected to each other and forming a wavy flank line. A cruciform or trident shaped mark on crown. Whitish below, uniform or spotted with brown.

HABITAT, DISTRIBUTION, and STATUS. Mainly inhabits arid country. Africa north of the equator, Middle East, India south and west of the Ganges. Not recorded in the Cochin and Travancore areas of Kerala, Sri Lanka. Common.

HABITS. Though essentially a desert snake, it occurs in semi-desert and broken scrub country. In some parts of the Deccan it has been recorded as extremely abundant over 2,25,000 being collected on an average per year for bounty over a period of six years in the Ratnagiri Dt. in the early years of the century. When the bounty payment was increased from six (3 p) pies to two annas (12 p) 1,15,921 were brought for collection in eight days!

Often seen basking in the full force of the sun or may retire below or into clefts of rocks too hot to touch from the sun's heat. In such conditions relies on moisture of the animals it eats but drinks when water is available. An alert little snake, it is largely diurnal and is capable of quick movement when necessary. In sandy areas it "side winds". Hibernates in the winter in its northern range but may emerge to bask in the heat of the sun. Often climbs on to shrubs and other low vegetation. The readiness with which it bites on the smallest provocation and the extremely fast strike makes it a very dangerous reptile. The striking posture is characteristic, a double coil in the form of a figure of 8 with its head in the centre. The coils keep moving against each other and the serrated keels on the flank scales produce a hissing noise by friction, amplified by the inflated body acting as a resonator. There is a report of a pot full of these snakes hissing under excitement sounding like a boiling kettle.

FOOD. Feeds largely on centipedes, scorpions, larger insects, mice, skinks, geckos and frogs.

BREEDING. The Saw-scaled Viper is viviparous producing 3 to 15 young at a time. Mating is believed to take place in the cold weather in north India, and young are born from April to August and occasionally in other months of the year. The young at birth measure 115 to 152 mm.

POISON. The fangs of this snake are remarkably long for its size, specimens 380 mm in length having 5 mm long fangs. The almond shaped poison glands are placed behind the eye. The average yield by weight of dry venon is about 18 mgm with a recorded maximum 72 mgm. About 12 mgm are injected at a bite roughly twice the lethal dose for an adult. The venom is said to be five times as toxic as cobra venom and sixteen times as toxic as Russell's viper venom and is very rapidly absorbed. Lethal dose for man is believed to be 5 mgm. The yield from snakes varies considerably and about 20% of bites may prove fatal.

The poison acts mainly as an anti-coagulent, a destroyer of blood cells and lining of blood vessels, a cardiac depressor and generally as a depressor to nerve cells. The local symptoms are similar to those of the Russells viper. The heart is strongly affected through the vasomotor centre in the brain resulting in weak pulse and low blood pressure. The venom acts directly on the cardiac muscle also. The blood cells are destroyed and haemorrhages almost inevitably occur from damaged blood vessels. Death results from heart failure. Death may occur within 24 hrs or less or when caused by exhaustion from repeated haemorrhages may occur after a week or two. The Haffekine Institute Polyvalent Serum is an effective antidote to the poison.

It is essential to keep the person under medical observation as delayed effects of the poison prove fatal even many days after the bite. The kidneys are often damaged beyond repair.

Green or Bamboo Pit Viper

Trimeresurus gramineus (Shaw)

HARMFUL

LOCAL NAMES. Hindi *Sakra*; Marathi *Hara ghonas*; Tamil *Pachai viriyan.*

SIZE. Average size 600 to 760 mm. Longest measured 1117 mm.

IDENTIFICATION. The loreal pit separates the pit vipers from all other groups of snakes. From other green pit vipers it is distinguished in having 15 to 17 rows of costals two headlengths before vent. The head is flattened and appears unduly broad owing to the constricted neck. Body stout. Taii short and tapering. Prehensile. Females usually have shorter tail.

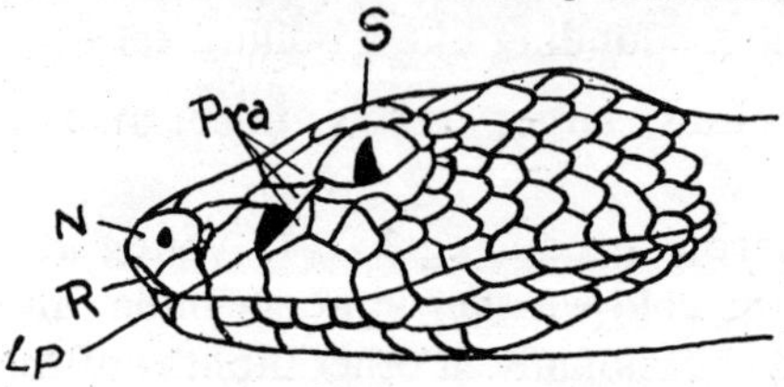

Head Scalation *Trimeresurus gramineus*

Key Character. Loreal pit (Lp)

COLORATION. Grass green above, glossy white, yellow or greenish below and on upper lip, chin and throat. A well-defined white, bluish or yellow flank line usually present. Tail yellowish or reddish mottled darker. Iris golden. Black or blackish markings occasionally on head and back. Markings on back may be in the form of crossbars. Occasionally yellowish or olivaceous rather than green.

HABITAT, DISTRIBUTION, and STATUS. A snake of hill forests. The snake is said to be confined to the Indian peninsula south of Lat. 22°. Not uncommon.

HABITS. A hill species not normally seen below 450 m. In the Western Ghats they are usually seen during the rains though rarely after October. Frequents low vegetation showing a marked preference for bamboo in localities where it occurs. Usually sluggish during the day as it reclines on branches four to eight feet from the ground its colour merging with its surroundings.

When provoked to strike, anchors itself firmly by its tail and hind body to a branch and with the rest of the body in a broad S, strikes with open mouth as far as it can straighten itself. The poison's action is feeble but painful consisting of pain and swelling of the bitten part, nausea, vomiting and fever the symptoms disappearing in about 48 hrs.

FOOD. Preferred food is small mammals but takes other small vertebrates also. Known to take small rats, mice, shrews, small birds, and lizards. A snake is also reported to have been taken. The tail tip is vibrated when about to strike at prey.

BREEDING. Viviparous producing 7 to 15 young at a time. Most births occur in June, July. Tip of tail of young reddish in colour.

Russell's Viper (*Vipera russelli*) (See p. 121)

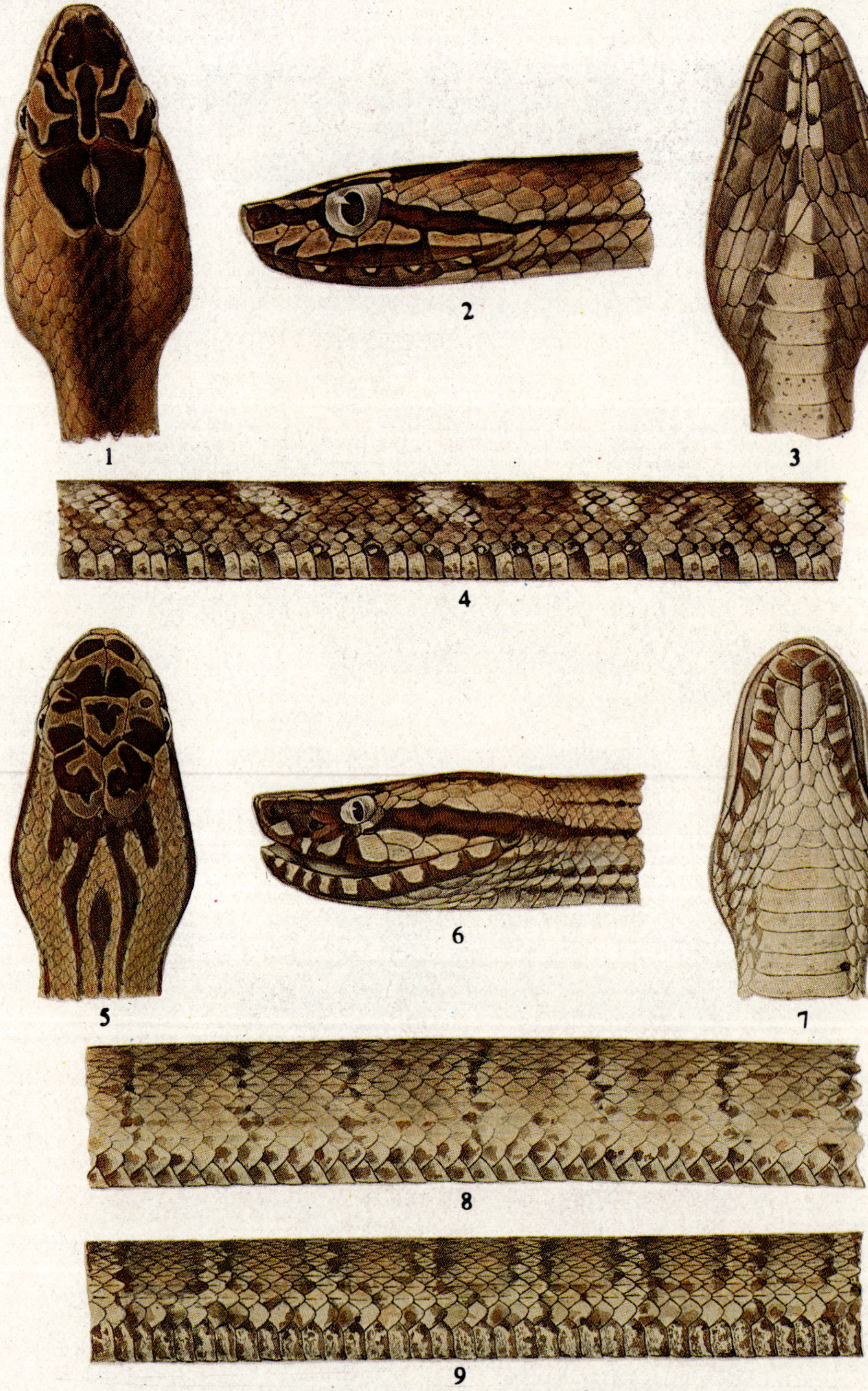

1-4. Mock Viper (*Psammodynastes pulverulentus*);
5-9. Himalayan Pit Viper (*Agkistrodon himalayanus*) (See p. 127)

HIMALAYAN PIT VIPER
Agkistrodon himalayanus (Günther)

HARMFUL

LOCAL NAME. *Pohur* Kashmir.

SIZE. Largest measured 34 in. (864 mm). Usually 18 to 24 in. (45.72 to 60.96 cm.).

IDENTIFICATION. The loreal pit identifies it as a pit viper; from other pit vipers by the presence of large shields instead of uniform small scales on top of head.

Body stout tapering to the anus, neck well marked. Head long and remarkably flat above. Eyes large with vertical pupil and gold flecked iris. Tail short with an elongate terminal shield. Dorsal scales dull and strongly ridged.

COLORATION. Variable. Brown of various shades up to blackish brown with irregular mottling or blotches of darker colour. Short crossbars occasionally present. Often a light vertebral line bordered with dark zigzag or sinous lateral stripes. Flanks dappled with shades of dark brown. White mottlings on side of neck not infrequent. Head usually darker than the back. Lips, chin and throat enamel white or pale pink. Ventrally plumbeous or dirty white powdered with various tones of brown, sepia or rufous. Tail tip usually reddish.

HABITAT, DISTRIBUTION, and STATUS. Usually occurs between 7000 and 10,000 feet (2133 to 3048 m.); rarer at higher altitudes though it has been recorded up to 16,000 ft. (4876 m) Prefers forested areas. Occurs commonly in the western Himalayas west of Nepal. Uncommon or rare east of Nepal.

HABITS. Essentially a forest lover and hardly ever seen away from vegetation. Usually takes refuge under logs, stones, walls of terraced fields and gardens, where it is common; occasionally enters tents of trekkers. A sluggish snake often seen basking close to cover on warm summer days. Hibernates for a lengthy period each year retiring in late autumn appearing again in spring. In keeping with its lethargic habits it is slow to anger, but if sufficiently annoyed will bite.

FOOD. Said to feed principally on mice. In Kashmir the main prey is the little skink *Lygosoma himalayanus*.

BREEDING. Viviparous. The mating season is not known but probably occurs during hibernation. Young 5 to 7 in one brood are born in August-September.

The poison is not particularly virulent and the bite of the snake is not fatal to man.

The MOCK VIPER *Psammodynastes pulverulentus* (Boie) though of an entirely different family bears a remarkable resemblance to the Himalayan Pit Viper *Agkistrodon himalayanus* in its colour and pattern — short and stout body, short tail, flattened head, large eye with vertical pupil, dull dorsal scales, attitude of menance and viviparous habit. However, it can be immediately separated from the viper by the absence of the loreal pit.

Curiously its distribution hardly overlaps that of the Himalayan Pit Viper as it is limited to the eastern Himalayas where it is not uncommon in forests between 3000 and 6000 ft(914 to 1828 m). Diurnal in habits, it is a lively snake which will strike if annoyed. Feeds mainly on reptiles. Ovo-viviparous; up to 10 young are born at a time in the summer months.

HUMP-NOSED PIT-VIPER

Hypnale hypnale Merrem

HARMFUL

A pit viper of the hill forests of South India and Sri Lanka distinguished by the enlarged scales on the head and the costals being in 17 rows at mid body. Grey or brown above, heavily powdered or mottled with dark brown and with large ovate, dark brown lateral spots, and red tail. Viviparous. The young have been noticed to have the remarkable habit of enticing skinks and other prey within reach by protruding the whitish tail from their coils and wriggling it around like a worm. Longest recorded 500 mm.

LITERATURE CITED

1. ABDULALI, HUMAYUN (1938): The food of the Mugger (*Crocodylus palustris*). *J. Bombay nat. Hist. Soc.* 40:336

2. ______ (1960): Notes on the spinytailed lizard, *Uromastix hardwicki* Gray. ibid. 57:421-423

3. ______ (1935): A dhaman (*Ptyas mucosus*) 'rattling' its tail. ibid. 37:958

4. ______ (1941): Rat snakes fighting. ibid. 42:666

5. ABERCROMBY, A. F. (1922): Crocodile (*C. palustris*) burying its food. ibid. 28:553

6. ALI, SALIM (1944): The 'courtship' of the monitor lizard (*Varanus monitor*). ibid. 44:479-480

6A. ANNANDALE, N. & SHASTRI, M.H. (1918): Relics of the worship of mud-turtles (Trionychidae) in India and Burma. *J. & Proc. Asiatic Soc.* Bengal NS 10:131-138

6B. ANNANDALE, N. (1915): Notes on some Indian Chelonia. *Rec. Indian. Mus.* 11:189

6C. ______ (1912): The Indian Mud-turtles (Trionychidae). ibid. 7:151-180

7. ASANA, J.J. (1931): The natural History of *Calotes versicolor* (Boulenger), the Common Bloodsucker. *J. Bombay nat. Hist. Soc.* 34:1041-1047

8. ______ (1941): Further observations on the egg laying habits of the lizard, *Calotes versicolor* (Boulenger). ibid. 42:937-940

9. AUFFENBERG, WALTER (1981): Combat behaviour in *Varanus bengalensis* (Sauria: Varanidae). ibid. 78:54-72

10. BANNERMAN, W.B. (1906): Note on the digestion of eggs by cobras and daboias. ibid. 16:363

11. BATTYE, R.K.M. (1944): Crocodiles Bellowing. ibid. 45:93-94

12. BHASKAR, SATISH (1979): Sea Turtle survey in the Andamans and Nicobars. *Hamadryad* 437:1-26

13. BROWN, W.C. & ALCALA, A.C. (1957): Viability of lizard eggs exposed to sea water. *Copeia* 1957:39-41

14. BURTON, R.W. (1950): The record Hamadryad or King Cobra (*Naja hannah* Cantor) and length and weights of large specimens *J. Bombay nat. Hist. Soc.* 49:561-562

15. BUSTARD, H.R. & BASU, D. (1982): A record (?) Gharial clutch. ibid. 79:207-208

16. CAMERON, T.H. (1923): Notes on turtles. ibid. 29:299-300

17. CHARI, V.K. (1950): A Dhaman or Rat Snake (*Ptyas mucosus*) (Linn.) jumping. ibid. 49:561

17A. CHAUDHURI, B.L. (1912): Aquatic tortoises of the middle Ganges and Brahmaputra. *Rec. Indian Mus.* 7:212-214

18. CHOPRA, R.N. (1964): Observations on the egg-laying of the Fanthroated Lizard, *Sitana ponticeriana* Cuv. *J. Bombay nat. Hist. Soc.* 61:190-191

19. CHOUDHURY, D.C. & BUSTARD, H.R. (1979): Predation on natural nests of the Saltwater Crocodile (*Crocodylus porosus* Schn.) on North Andaman Island with notes on the crocodile population. ibid. 76:311-323

20. D'ABREU, E.A. (1913): Effect of a bite from Schneider's Water-snake (*Hypsirhina enhydris*). ibid. 22:203

21. ______ (1932): Notes on Monitor Lizard. ibid. 36:269-270

22. DANIEL, J.C. & HUSSAIN, S.A. (1975): The record (?) Saltwater Crocodile (*Crocodylus porosus* Schneider). ibid. 71:309-312

22A. ______ & SHULL, E.M. (1964): A list of the reptiles and amphibians of Surat Dangs. ibid. 60:737-743

23. DERANIYAGALA, P.E.P. (1958): Reproduction in the Monitor Lizard, *Varanus bengalensis* (Daudin). *Spolia Zeylanica* 28:161-166

24. DHARMAKUMARSINHJI, K.S. (1947): Mating and the parental instinct of the Marsh Crocodile (*C. palustris* Lesson). *J. Bombay nat. Hist. Soc.* 47:174-176

25. DODSWORTH, P.T.L. (1911): The food of crocodiles. ibid. 20:523

26. DUNBAR, BRANDER (1923): Wild Animals of Central India. Edward Arnold & Co. pp. 284-288

27. ECKARDT, M.J. (1971): Skin homografts in the all female Gekkonid Lizard, *Hemidactylus garnoti. Copeia* 1971:152-154

28. EVANS, G.H. (1906): Breeding of the Banded Krait (*Bungarus fasciatus*) in Burma. *J. Bombay nat. Hist. Soc.* 16:519-520

29. FORSYTH, H.W. (1912): The food of Crocodiles. ibid. 20:228

30. FRASER, A.G.L. (1937): The Snakes of Deolali. ibid. 39:464-501

31. GHARPUREY, K.G. (1931): An unusually large Shaw's Rat Snake (*Zamenis fasciolatus*). ibid. 34:1084

32. GHOSH, S.K. (1948): A King Cobra's speed. ibid. 47:760-761

33. HAAST, W.E. & WERNER, M.L. (1953): The second most poisonous snake in the World: *Amer. J. of Trop. Med. & Hygiene* Vol. 4, No. 6:1135-1137

34. HAGENBECK, JOHN (1906): Size and breeding of snakes. *J. Bombay nat. Hist. Soc.* 16:505-506

35. HARRISSON, TOM (1959): Notes on the edible green Turtle (*Chelonia mydas*). *Sarawak Museum Journal* 9:277-278

35A. HENDERSON, J.R. (1912): Preliminary note on a new Tortoise from south India. *Rec. Indian Mus.* 7:217-218

36. HENDRICKSON, J.R. (1958): The Green Sea Turtle, *Chelonia mydas* in Malaya and Sarawak. *Proc. zool. Soc. London* 130:455-535

37. HUTTON, ANGUS F. (1949): Notes on the snakes and mammals of the High Wavy mountains, Madura Dt., S. India. *J. Bombay nat. Hist. Soc.* 48:454-460

38. JASDAN, RAJA of (1953): Python capturing Chinkara. ibid. 51:945-946

39. JOHN, K.O. (1962): Notes on the bionomics of the Flying Lizard *Draco dussumieri* Dum. & Bib. ibid. 59:298-301

40. ______ (1967): Observations on the mating behaviour and copulation in *Draco dussumieri* Dum. & Bib. (Reptilia: Sauria). ibid. 64: 112-115

41. JONES, S. (1959): A leathery turtle *Dermochelys coriacea* (Linnaeus) coming ashore for laying eggs during the day. ibid. 56:137-139

42. KAUFFELD, C.F. (1954): Manipulation of odour as an aid in feeding captive snakes with special reference to King Cobras. *Herpetologica* 10:108-110

43. KEAYS, R.W. (1929): An unpleasant experience with a python. ibid. 33:721-722

44. KLUGE, A.G. & ECKARDT, M.J. (1969): *Hemidactylus garnoti* Dum. & Bibr., a triploid, all female species of Gekkonid Lizard. *Copeia*: 651-64

45. KAROPACH, CHAIM (1971): Sea Snake (*Pelamus platurus*) aggregation on slicks in Panama. *Herpetologica* 27:131-135

46. KUNTZ, R.E. (1963): Snakes of Taiwan. *Quart. J. Taiwan Mus.* 16:1-79

47. KURIYEN, G.K. (1950): Turtle fishing in the sea around Krusadi Island. *J. Bombay nat. Hist. Soc.* 49:509-512

48. LAHIRI, R.K. (1955): A 'white python'. ibid. 53:135-136

49. LEIGH, C.J. (1949): Egg laying by the Indian python in captivity. ibid. 48:597

50. LEVER, J. (1975): Behaviour of *Crocodylus porosus* —Defence of nest. Privately circulated information.

51. MACDONALD, A.S.T.J. (1947): 'Shamming Death'—Snakes. *J. Bombay nat. Hist. Soc.* 47:173

52. MCARTHUR, A.G. (1922): A python's long fast. ibid. 28:1142-1143

53. MCCANN, C. (1935): Male Rat-Snakes (*Zamenis mucosus* Boulenger) fighting. ibid. 38:409

54. ______ (1938): The reptiles and Amphibia of Cutch State. ibid. 40:425-427

55. MCCANN, C. (1940): A reptile and amphibian miscellany. *J. Bombay nat. Hist. Soc.* 41:742-764

56. MINTON, S.A. (1966). A contribution to the herpetology of W. Pakistan. *Bull. Amer. Mus. nat. Hist.* 134:27-184

57. MROSOVSKY, N. (1968): Nocturnal emergence of hatchling sea turtles control by thermal inhibition of activity. *Nature* 220., No. 5124, pp. 1338-1339

58. MORRIS, R.C. (1958): Rat snakes 'mating'. *J. Bombay nat. Hist. Soc.* 55:366

59. MOSES, S.T. (1948): Crocodiles in India. *Bulletin* 15, Dept. of Fisheries, Baroda

59B. PRASHAD, BAINI (1914): Notes on aquatic Chelonia of the Indus system. *Rec. Indian Mus.* 10:267-271

59C. ______ (1914): Lizards of the Simla Hill States. ibid. 10:367-369

60. ______ (1916): Some observations on a common house lizard (*Hemidactylus flaviviridis* Ruppell) of India. *J. Bombay nat. Hist. Soc.* 24:834-838

60A. PRITCHARD, PETER C.H. (1979): Encyclopedia of Turtles. T.F.H. Publications, Neptune, N.J., USA

61. PURVES, E. HOME (1915): The Thorny Tailed Lizard. *J. Bombay nat. Hist. Soc.* 23:780-784

62. RICHARD, B.D. (1917): Note on the habits of the Checkered Water Snake (*Tropidonotus piscator*). ibid. 25:150

63. RUSSELL, A.P. (1975): A contribution to the functional analysis of the foot of the Tokay, *Gekko gecko* (Reptilia: Gekkonidae). *J. zool. London* 176:437-476

64. SANJEEVA RAJ, P.J. (1958): Egg laying habits of sea turtles described in the Tamil Sangam Literature. *J. Bombay nat. Hist. Soc.* 55:361-363

65. SANYAL, M.K. & PRASAD, M.R.N. (1967): Reproductive cycle of the Indian House Lizard *Hemidactylus flaviviridis* Ruppell. *Copeia* 1967:627-633

66. SETHNA, K.R. (1958): Rat-Snakes 'mating'. *J. Bombay nat. Hist. Soc.* 55:173-174

67. SIMCOX, A.H.A. (1906): The crocodile, its food and muscular vitality. ibid. 16:375-376

68. SIMON, E.S. (1944): The breeding habits of the Cobra (*Naia tripudians* Merrem) and the Green Whip snake (*Dryophis mycterizans*). ibid. 44:480-481

69. SINGH, H.K. LALA (1980): Sexual attraction of a wild Mugger (*Crocodylus palustris* Lesson) towards captive muggers. ibid. 76:167-171

70. SINGH, V.B. (1979): The status of Gharial in U.P. and its rehabilitation. ibid. 75:668-683

71. SMITH, H.C. (1930): The Monitor Lizards of Burma. ibid. 34:367-373

72. SUBRAHMANIAM, T.V. (1934): Rat snakes and their food value. *J. Bombay nat. Hist. Soc.* 37:743

73. TIKADER, B.K. (1968): Observations on the Limbless Lizard *Ophisaurus gracilis* (Gray) from Shillong, Assam. ibid. 65:233

74. TIWARI, K.K. (1961): The eggs and flight of the Gecko *Ptychozoon khuli* Stejneger from Car Nicobar. ibid. 58-523-527

75. TRENCH, CHENEVIX C. (1912): Notes on the Indian Chamaeleon (*Chamaeleon calcaratus*). ibid. 21:687-689

76. TWEEDIE, M.W.F. (1954): Notes on Malayan reptiles No. 3. *Bull. Raff. Mus.* No. 25:107-117

77. UNDERWOOD, GARTH (1948): Notes on Poona Reptiles. *J. Bombay nat. Hist. Soc.* 47:627-632

77A. VIJAYA, J. (1982): Successful artificial breeding of *Lissemys punctata granosa.* ibid. 79:210-11

77B. ______ (1982): Rediscovery of the forest cane turtle *Heosemys (Geoemyda) silvatica* (Reptilia, Testudinata, Emydidae) from Chalakudy forests in Kerala. ibid. 79(3):676-677

77C. ______ (1983): Worlds rarest turtle lays eggs in captivity—*Hamadryad* (newsletter of the Madras Snake Park) 8(1):13

78. WALL, F. (1906): Notes on snakes collected in Cannanore from 5th November 1903 to 5th August 1904. *J. Bombay nat. Hist. Soc.* 16:292-312

79. ______ (1906): The breeding of Russell's Viper (*Vipera russelli*). ibid. 16:374-375

80. ______ (1908): Notes on the incubation and brood of the Indo-Burmese Snake-Lizard or Slow Worm (*Ophisaurus gracillis*). ibid. 18: 503-504

81. ______ (1909):Notes on snakes collected in Upper Assam. ibid19:608623

82. WEBB, GRAHAME (1977): The natural history of *Crocodylus porosus* in "Australian animals and their environment". Ed. by Messel & Butler, Shakespeare Head Press, Sydney.

83. WHITAKER, R. (1978): Breeding record of the Indian Chameleon (*Chameleo zeylanicus*). *J. Bombay nat. Hist. Soc.* 75:232

84. ______ (1978): Note on the status of the Gir crocodiles. ibid. 75:224-227

85. ______ (1978): A preliminary survey of the Salt water Crocodile (*Crocodylus porosus*) in the Andaman Islands. ibid. 75:43-49

86. ______ & WHITAKER, Z. (1978): Growth rate of *Crocodylus palustris.* ibid. 75:231-232

87. WHITAKER, R. & WHITAKER, Z. (1978): Notes on captive breeding in Mugger (*Crocodylus palustris*). *J. Bombay nat. Hist. Soc.* 75:228-231

88. ______ (1979): Breeding of Tokay gecko. ibid. 75:499

89. ______ (1979): Notes on *Phelsuma andamanense* the Andaman day gecko or green gecko. ibid. 75:497-499

90. WILLIAMSON, H.A. (1967): Notes on growth rate of *P. reticulatus. Herpetologica*: 30

91. WRENICKE, C.J.T. (1955): Pythons. *J. Bombay nat. Hist. Soc.* 53:134-135

92. ZELLER, WARREN (1969): Maintenance of the Yellow-bellied Sea-snake *Pelamis platurus* in captivity. *Copeia* 1969

93. ZUG, GEORGE R. (1974): Crocodilian galloping: A unique gait for reptiles. ibid. 1974:550-552

ADDITIONAL READING

BELLAIRS, ANGUS (1969): THE LIFE OF REPTILES, vols. 1 and 2. London. Weidenfeld and Nicolson.

BELLAIRS, A. D'A. & CARRINGTON, R. (1966): THE WORLD OF REPTILES. London. Chatto & Windus; N. Y., American Elsevier.

BUSTARD, ROBERT (1972): SEA TURTLES. NATURAL HISTORY AND CONSERVATION. London. Collins.

CARR, A. (1968): THE TURTLE. A NATURAL HISTORY. London. Cassell; (1967) N.Y., Natural History Press.

DITMARS, R.L. (1937): SNAKES OF THE WORLD. New York. Macmillan. (1st Publ. 1931).

GANS, C., BELLAIRS, A. D'A. & PARSON, T.S. (Eds) (1969): BIOLOGY OF THE REPTILIA, Vol. 1. New York & London. Academic Press (A multi-author and multi-volume compilation).

GOIN, C.J., GOIN, O.B. & ZUG, G.R. (1979): INTRODUCTION TO HERPETOLOGY (3rd ed.). San Francisco. W.H. Freeman & Co.

PARKER, H.W. (1965): NATURAL HISTORY OF SNAKES. London. Brit. Mus. (Nat. Hist.)

POPE, C.H. (1956): THE REPTILE WORLD. A NATURAL HISTORY OF THE SNAKES, LIZARDS, TURTLES AND CROCODILIANS. London. Routledge & Kegan Paul; N.Y., Knopf.

PRITCHARD, PETER C.H. (1979): ENCYCLOPEDIA OF TURTLES. Neptune, N.J. T.F.H. Publications Inc.

SMITH, M.A. (1931-43): THE FAUNA OF BRITISH INDIA, CEYLON AND BURMA, INCLUDING THE WHOLE OF THE INDO-CHINESE SUB-REGION. REPTILIA AND AMPHIBIA (The Amphibia are not included). 3 Vols. London. Taylor & Francis.

WHITAKER, ROMULUS (1978): COMMON INDIAN SNAKES. New Delhi. Macmillan Co. of India Ltd.

ABBREVIATIONS

ag — anterior genials (chin shields)
A.S. — anterior sublingual
f — femoral
in — internasal
l — loreal
la' — infralabial
la — supralabial
m — mental
n — nasal
p, pa — parietal
pf — prefrontal
pg — posterior genial
po — postocular
pr, pro — preocular
prs — presubocular
P.S. — posterior sublingual
t — anterior and posterior temporals
v — ventrals

GLOSSARY

Aestivation — prolonged summer torpor
Alveolar — pitted
Axillary — of the angle between forelimb and body
Carapace — Bony upper shield of a tortoise or turtle
Caudal — of the tail
Caul — enclosing membrane
Cloaca — chamber into which open the anus, urinary duct and genital ducts
Columnar — vertically elongated
Costal — rib or upper body shields
Crepuscular — active at twilight or preceding dawn
Cryptic — hidden
Cuneate — wedge shaped
Femoral — of the thigh
Frontal — head shield between orbits
Haemotoxin — blood poison
Herbivorous — feeding on vegetable matter
Hibernation — complete or partial torpor in winter
Imbricate — overlapping like tiles on a roof
Inguinal — groin area
Internasal — shield between nasal shields
Keel — elongated ridge on scale
Labial — of the lip
Lamellae — thin plates
Loreal — space between the snout and the eye
Marginal — plates forming the edge of the carapace
Matrix — intercellular ground connective tissue
Mental — of the chin
Necrosis — death of cell/cells while still part of the living body
Nuchal — of the back of the neck
Olfactory — pertaining to the sense of smell
Omnivorous — feeding on both animal and vegetable matter

Ovoviviparous — producing eggs which hatch within the uterus of the mother
Parthenogenesis — reproduction without fertilisation by the male
Patagium — membranous wing-like structure
Phalanges — digit bones
Phrenic — pertaining to the diaphram
Placebo — A substance having no medicinal value but given as a placative
Plastron — bony ventral shield of a tortoise or turtle shell
Plantigrade — walking on the soles of the feet
Polyvalent — carrying immunity against more than one venom
Prefrontal — head shield preceding frontal shield
Preocular — in front of eye
Proteolytic — having enzymes which break down proteins into simpler substances
Rostral — of the snout
Scutes — scales or plates
Setae — small bristle like structures
Sinus — cavity or depression of irregular size
Supracaudal — above the tail
Supralabial — of the upper lip
Supranasal — above the nostril
Temporal — of the temple region
Triploid chromosomes — having thrice the germ cell chrosome number
Ventrals — belly shields
Vertebrals — of the spine
Viviparous — giving birth to live young

INDEX

ERRATA

Page 37 and plate facing p. 36
For *Gecko gecko* read *Gekko gecko.*

Page 56 and plate facing p. 43
For *Opisops* read *Ophisops.*